KU-825-851

156.95
T

This book is to be returned on or before
the last date stamped below.

2 JAN 199   17 APR 2013

26 NOV 1991

16 DEC 1991

5 FEB

25 JUN 1998

1998

7 DEC 1998

17 DEC 2008

1 APR 2010

LIBREX —

E00247

A Colour Atlas of
Gynaecological Cytology

A Colour Atlas of
GYNAECOLOGICAL CYTOLOGY

O. A. N. Husain
MD, FRCPath, FRCOG
Consultant Pathologist
Charing Cross Hospital and
St Stephen's Hospital
London

E. Blanche Butler
MD, FRCOG, MRCPath
Early Diagnostic Unit
Elizabeth Garrett Anderson Hospital
London

Wolfe Medical Publications Ltd

Copyright © O. Husain, E. Butler, 1989
First published 1989 by Wolfe Publishing Ltd
Printed by W.S. Cowell Ltd, Ipswich, England
ISBN 0 7234 0913 7

A CIP catalogue record for this book is available from the British Library.

For a full list of Wolfe Medical Atlases, plus forthcoming titles and
details of our surgical, dental and veterinary Atlases, please write to
Wolfe Publishing Ltd, 2-16 Torrington Place, London
WC1E 7LT, England.

Contents

Preface

This atlas has been written with both pathologists and gynaecologists firmly in mind. Both need to identify the varying shades of cellular changes occurring in the lower gynaecological tract. Many pathologists look on gynaecological cytology simply as a screening technique used to identify women with cervical smear abnormalities who may warrant further investigation—such as colposcopy and biopsy—so that a definitive diagnosis may be made. Recent developments in the fields of endoscopic cytology and fine-needle aspiration have interested pathologists more and many have achieved a diagnostic expertise which they may not apply when examining cervical smears. As a result there is a danger of underestimating the cytologist's ability to achieve a high quality diagnostic result.

The object of this atlas is to illustrate the wide range of cytological appearances in cervical smears which can lead to a positive and reliable diagnosis in gynaecological disease generally, and neoplasia in particular. Special emphasis is given to problem areas and cases where there are difficulties in differential diagnosis.

The material presented has been collected over many years in the laboratories of the authors (from St Stephen's Hospital, London, Charing Cross Hospital, London and St Mary's Hospital, Manchester) and a few from colleagues, which are acknowledged. With this diversity of provenance and variation of preparations the quality of the photographs is likely to be uneven. We have included specimens that are not ideal for photographic presentation in order to give the reader the greatest range of material seen in routine practice.

Diagnosis in gynaecological cytology is closely related to the complementary disciplines of colposcopy and histopathology, so relevant examples are included. In the case of colposcopy and mounted whole sections of tissue blocks it is hoped that this will help in understanding the disease, while the examination of histological sections is essential for cell correlation.

We hope that the atlas will provide a useful bench book for pathologists, a valuable lesson on the value of cytology for the gynaecologist, and also be a useful instructional and visual guide for students of cytopathology. As an atlas of cytological appearances it cannot provide a detailed text or bibliography; for this readers are referred to standard textbooks.

Acknowledgements

We are grateful to Professor V. R. Tindall and Wolfe Publishing for permission to reproduce the following colposcopic pictures: Figures 1, 22, 139, 140, 196, 240, 327, 411 and 417. Figures 244, 245, 262 and 271 were photographed at Johns Hopkins Hospital and are included with the kind permission of Dr J. K. Frost. Dr Elsa Skaarland has been kind enough to allow us to include photographs of her material to illustrate the section on endometrial aspiration cytology (Figures 304-311). Figures 304, 307 and 311 were taken by Dr Skaarland.

Other material has been provided by: Dr Aileen Hampton (Figure 151), Dr C. M. Stanbridge & Dr C. H. Buckley (Figures 426-433), Dr Hilda Harris (Figure 353), Dr Andrew Kertesz (Figures 246-248), The Edgware General Hospital (Figures 256-258), Mr Reddish (Figure 125) and Dr Elizabeth Benjamin (Figures 377, 380, 384, 385, 390, 392, 395 and 398). Dr Benjamin also reviewed these sections.

Staining and magnification

Unless otherwise stated, the material photographed is taken from cervical smears and the staining is by modifications of the Papanicolaou method. The magnifications given are the original frame magnifications. These are reproduced at approximately 250 per cent of the original frame size.

1   Cell content of normal cervical smears

Cervical smears should be taken from the squamocolumnar junction so as to include cells from both squamous and columnar epithelium. It is essential that the transformation zone is also included in the 360° scrape covered by the spatula. Sampling errors are an important cause of false negative smears, therefore the technique of taking smears should be perfected as much as possible (*Macgregor, 1981*). Speed of fixation is important, and it is preferable to use 74° OP alcohol to which polyethylene glycol has been added. This provides a hygroscopic cover so that fixed slides are allowed to dry before they are sent for staining and examination. Fixatives can be used in drip or spray form. For routine work, Papanicolaou's stain is used on cervical smears.

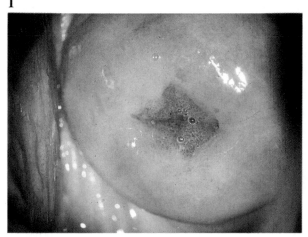

1 Normal cervix, colposcopy. This shows squamous epithelium covering the ectocervix, and vascular columnar epithelium lining the endocervical canal (see also **14**). Here the columnar epithelium is thrown into fine villi giving a 'bunch of grapes' appearance. The bivalve speculum used to expose the cervix also causes slight eversion of the external os and opens up the endocervical canal. (*Saline, × 16*)

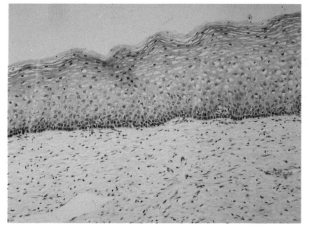

2 Cervix: normal stratified squamous epithelium. The germinal layer is seen at the base of the epithelium. There is regular maturation of the cells through the parabasal layers to the intermediate and superficial cell layers. Unlike skin, squamous epithelium covering the cervix and vagina does not normally develop a keratinized layer. (*H&E, x 40*)

Variation of squamous cell exfoliation

In cervical smears, squamous cells are obtained from the surface of the epithelium, therefore in a sexually mature woman these will be superficial and intermediate cells only. As the epithelium is under hormonal influence, the pattern of exfoliating cells varies depending on the stage of the menstrual cycle, during pregnancy or puerperium, and also in the prepubertal girl and the postmenopausal woman.

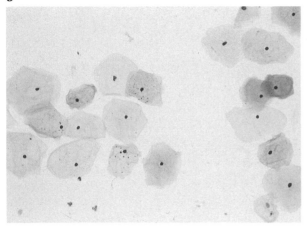

3 Oestrogen effect. This appearance is typical of a cervical smear taken at mid-cycle, when ovulation is imminent and an oestrogen effect predominates. The squames present are mainly superficial cells placed discretely with pyknotic nuclei and a flat, translucent cytoplasm. On the whole, the more mature cells have eosinophilic cytoplasm, but this is not as specific as with Shorr's stain and a range of staining reactions will be noted. Keratohyalin granules are seen in the cytoplasm of some cells. (× *80*)

4 Postovulatory smear. This field shows the effect of progesterone produced by the corpus luteum after ovulation. Superficial cells with pyknotic nuclei and keratohyaline granules are still seen (arrow), but there are increased numbers of intermediate squamous cells with vesicular nuclei showing a fine chromatin pattern. The cytoplasm of these cells is more likely to be cyanophilic, but it may be eosinophilic. It will be noted that there is some clumping of the cells with folding of the cytoplasm. If vaginal wall smears are taken daily, it is possible to identify the time of ovulation by the sudden change of pattern from that seen in **3** to that seen in **4**. However, estimation of hormone levels by modern methods has eliminated this investigation from cytology laboratories. (× *160*)

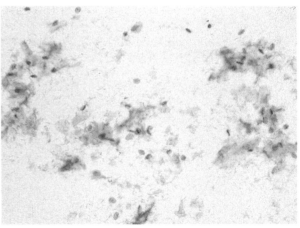

5 Pregnancy pattern. During pregnancy there is an exaggeration of the progesterone effect, with marked clumping of cells and folding of the cytoplasm. Intermediate cells predominate and few superficial cells are seen. Note the fraying of the cytoplasmic borders. (× *80*)

6 Pregnancy pattern. This field shows marked fraying and loss of cytoplasm, together with a marked bacterial haze of lactobacilli. The result is a cytolytic smear which is common in pregnancy and is also seen in women taking certain forms of oral contraception. (× *80*)

7 Pregnancy pattern. This field shows another feature seen in cervicovaginal smears in pregnancy. Cells show a folded cytoplasm and are moulded together, with rolled edges, to the cytoplasm. This results in the formation of characteristic shapes and hence the description of cells as 'oyster' and 'navicular' cells. In addition, nuclei may be ovoid and eccentrically placed. The yellow haze in some cells indicates the presence of glycogen. (× *160*)

8 Puerperal pattern. This field contains darkly staining parabasal cells and lighter, small, intermediate cells. Following delivery, the cell pattern may become atrophic (*Butler & Taylor, 1973*). Not all women develop this pattern; there is also variation in the length of time during which an atrophic cell pattern may be present. It has been suggested that this change is due to a sudden fall in hormone levels after delivery of the placenta, but there is also evidence to support the view that it is caused by regeneration of the epithelium following the trauma of delivery. (× *80*)

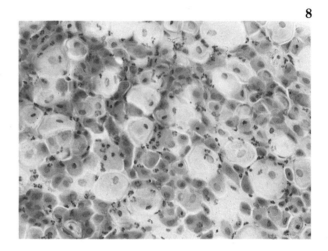

9 Puerperal pattern. This smear is taken at a later stage of the puerperium. It shows a mixed cell pattern with all cell types being present, and reflects the progress of tissue regeneration. (× *80*)

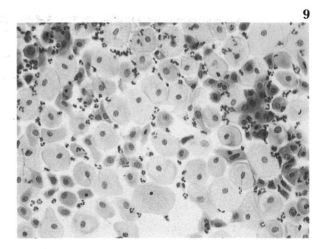

Postmenopausal smears

Following the menopause, three main types of cell pattern are seen: intermediate (or proliferative), atrophic and mixed. The cell pattern observed seems to bear no relationship to the presence or absence of systemic menopausal symptoms or to hormone levels in the blood. In addition, there is no evidence that this pattern is progressive, becoming more atrophic as the woman ages. It would seem to be an individual response of the target organ (vaginal epithelium) to the woman's steroid metabolism.

Superficial cells should be absent or scanty in postmenopausal smears. If more than 10 per cent are present, further investigation is needed to identify the reason for this apparent oestrogen effect, which may accompany tumours of the ovary, endometrium and breast. It also occurs when there has been long-term treatment with drugs such as digitalis and tranquillizers. However, a few cases of normal women with raised superficial cell counts have been reported (*DeWaard & Baanders-van Halewijin, 1969; DeWaard et al., 1972*).

10

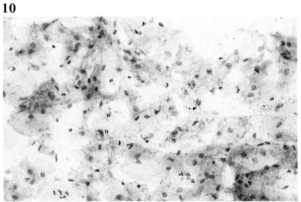

10 Postmenopausal smear: intermediate cell pattern. This is a postmenopausal smear showing an intermediate cell pattern. The epithelium is still relatively thick and matures to the intermediate cell level. (× *80*)

11

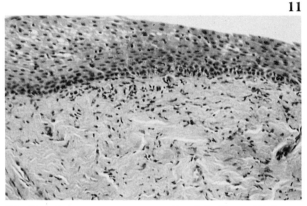

11 Cervix: postmenopausal epithelium. This section shows the more usual appearance of vaginal and cervical squamous epithelium. There is compaction of the cell layers, and maturation ceases at the parabasal or small intermediate cell level. (*H&E,* × *62*)

12

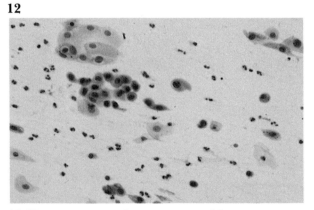

12 Postmenopausal smear: atrophic pattern. In this smear the cell pattern correlates with the epithelium seen in **11**. The cells present range from parabasal to small intermediate cells. There is some nuclear degeneration, and perinuclear halos are seen. Inflammatory cells are present due to some degree of vaginitis, common when the epithelium is atrophic. (× *80*)

13

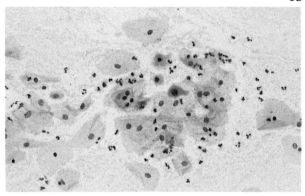

13 Postmenopausal smear: mixed cell pattern. The third common pattern is demonstrated in this field. It probably reflects a greater degree of atrophic vaginitis which stimulates regenerative changes in the epithelium, therefore all cell types are present. Note the heavy coccal haze and the presence of polymorphs. In some cases lymphocytes and/or histiocytes predominate; this suggests further damage to the epithelium resulting in a granulomatous reaction (see **96** and **97**). (× *80*)

Endocervical columnar cells

The endocervical canal is lined by a single layer of tall, mucus-secreting, columnar cells with basal nuclei. The lining epithelium in the canal is thrown into folds which may be coarse and longitudinal, or more finely grouped as papillary villi. This appearance is seen in **1**.

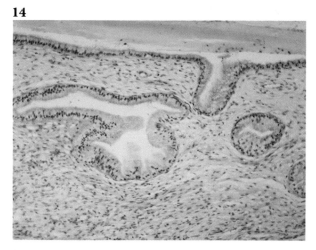

14 Normal endocervical columnar epithelium. In this section the superficial part of a crypt is seen indenting the surface, but because of the plane of section the remainder of the crypt looks like a glandular structure in the stroma. It is important to remember that the epithelium lining 'glands' in the endocervical stroma is continuous with the surface epithelium. (*H&E, × 40*)

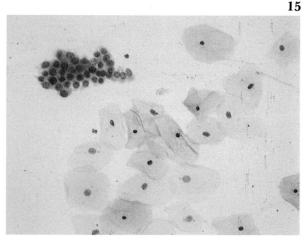

15 Endocervical columnar cells. Endocervical columnar cells may exfoliate as single cells (see **45**), but it is more usual for exfoliation to occur as tissue fragments. In this field most of the cells are seen *en fosse*, presenting a honeycomb effect, but at one edge three cells are seen in profile. (× *80*)

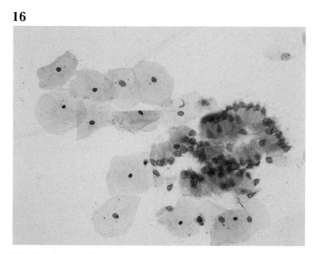

16 Endocervical columnar cells. In this field the strips of endocervical columnar cells are all seen in profile. They have basal nuclei with foamy cytoplasm. In some of the cells an end-plate and cilia can be recognized. (× *80*)

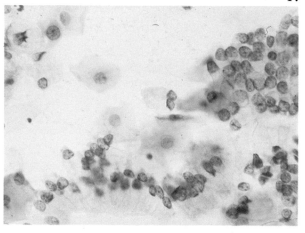

17 Endocervical columnar cells. At higher magnification it is possible to contrast the nuclear chromatin pattern of normal squamous cells with that of normal endocervical columnar cells. In the latter the chromatin is more granular, with condensation at the nuclear membrane. (× *160*)

Squamous metaplasia and the transformation zone

Under the influence of oestrogen the cervix becomes everted to expose the columnar epithelium of the canal. This is most marked at puberty and during pregnancy. It is also seen in women on high-oestrogen oral contraceptives. The acid pH of the vagina stimulates replacement of columnar epithelium with squamous epithelium by a process of squamous metaplasia (*Singer & Jordan, 1976*). The area in which this initially occurs is known as the 'transformation zone'. In the early stages of squamous metaplasia, undifferentiated cells multiply between the columnar cells and the basement membrane. These may originate in situ, but where there has been destruction of surface epithelium, for example by laser or diathermy, a similar layering of cells occurs. This suggests migration from the stroma, which may also be the case in physiological metaplasia. At a later stage the cells stratify and the epithelium is recognizably squamous, eventually maturing to be indistinguishable from original squamous epithelium. Although the epithelial layers look the same, the epithelium can be recognized as metaplastic because endocervical crypts are seen in the underlying stroma.

Groups of immature metaplastic cells can be identified in cervical smears. They are seen as parabasal cells in a field which otherwise consists of mature squamous cells. In these cases they are referred to as metaplastic cells. They may be associated with columnar cells, and their presence in a smear also indicates an acceptable specimen for reporting. The views in this section illustrate the process of squamous metaplasia and also show the range of appearances of metaplastic cells.

18

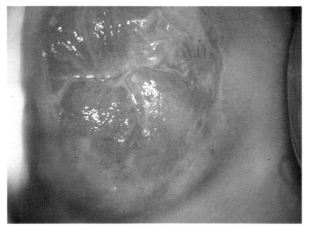

18 Ectopy with early transformation zone: colposcopy. Here the cervix is everted to show folds of vascular endocervical columnar epithelium. At the extreme periphery of the portio of the cervix, original smooth squamous epithelium is present. Between these areas and in patches on the ectopy there is immature metaplasia which turns white after the application of 4 per cent acetic acid. (*Acetic acid, × 16*)

19

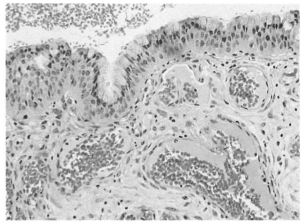

19 Cervix: reserve cell hyperplasia. This section shows the replication of indifferent cells deep to the columnar epithelium, which is called reserve cell hyperplasia. (*H&E, × 62*)

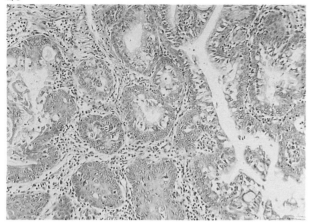

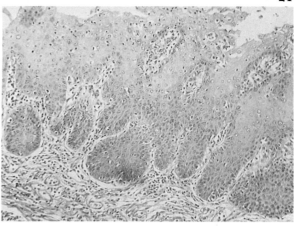

20 Cervix: reserve cell hyperplasia: gland involvement. In this field the branching crypts in the stroma are seen to be undergoing the same process of reserve cell hyperplasia. (*H&E, × 37.5*)

21 Cervix: immature squamous metaplasia. At a later stage the layered immature cells, both on the surface and in the crypts, are beginning to stratify and an immature squamous epithelium can be recognized. (*H&E, × 37.5*)

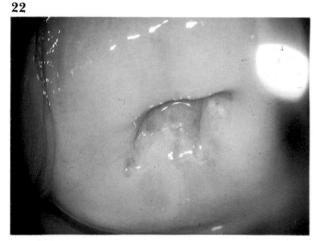

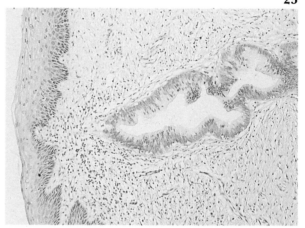

22 Healed transformation zone: colposcopy. Here the columnar epithelium of the ectopy is replaced by mature metaplastic epithelium. When viewed three-dimensionally through the colposcope, it is usually possible to see round indentations on the portio which identify the level of the original squamocolumnar junction. The new squamocolumnar junction is seen at the external os. Between original and new squamocolumnar junctions there is mature metaplastic squamous epithelium. (*Acetic acid, × 16*)

23 Mature metaplastic epithelium with underlying crypt. This field shows mature metaplastic squamous epithelium. Histologically it resembles original squamous epithelium and is distinguished by the crypt in the underlying stroma. (*H&E, × 40*)

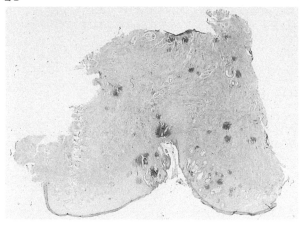

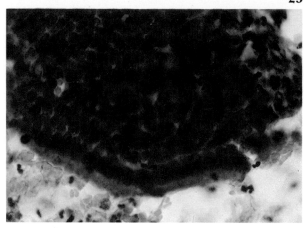

24 Section of cervix. This is a thick section of cervix. It demonstrates original squamous and columnar epithelium, with metaplastic epithelium between them. Note the crypts deep to the mature metaplastic epithelium, and also the depth of penetration of the crypts into the stroma. (*H&E, × 1*)

25 Endocervical columnar and reserve cells. This field shows a tissue fragment which consists of a strip of endocervical columnar cells and underlying reserve cells. (*× 160*)

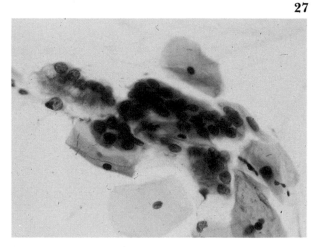

26 Metaplastic and reserve cells. Mature squamous cells are present, with endocervical columnar cells and clusters of reserve and immature metaplastic cells. Note the wide range of maturation. (*× 80*)

27 Endocervical columnar and reserve cells. Endocervical columnar cells are seen in profile, with vacuoles in the cytoplasm. The reserve cells are closely packed and show granular hyperchromatic nuclei. (*× 160*)

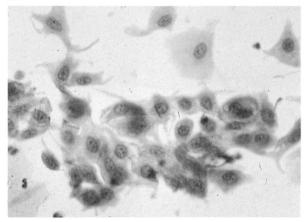

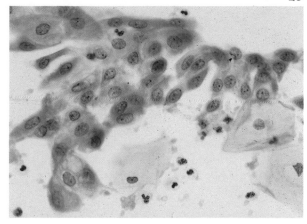

28 Immature metaplastic cells. These cells have more cytoplasm than reserve cells, but are still quite immature. Because of the cytoplasmic processes, they are called 'spider' cells. This appearance indicates that the cells have been removed forcibly by scraping the cervix, instead of being exfoliated cells lying on the surface (*Patten, 1978*). (× *125*)

29 Immature metaplasia. The presentation of metaplastic cells in cervical smears is very variable. In this field, cells are seen as a loosely connected sheet and the nuclei show prominent nucleoli. The appearances may reflect healing of a true ulcer or repair of trauma (see **99**). (× *160*)

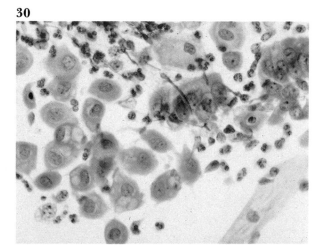

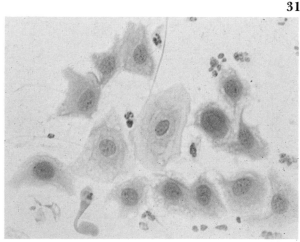

30 Endocervical columnar and metaplastic cells. In this smear there is more rounding of the cytoplasm of the metaplastic cells, and some show cytoplasmic vacuolization. There is a mixed inflammatory cell exudate with polymorphs and histiocytes, and endocervical columnar cells show prominent nucleoli. It seems probable that cervicitis, as well as a healing transformation zone is present. (× *160*)

31 Metaplastic cells. In this field a full range of metaplastic cells of varying maturity lie together, so that it is possible to compare the change of nuclear chromatin pattern and cytoplasmic differentiation from reserve cells to almost mature metaplastic cells of the intermediate cell type. (× *125*)

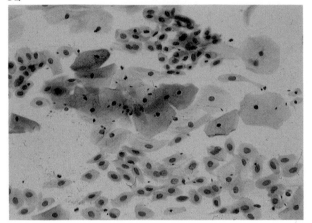

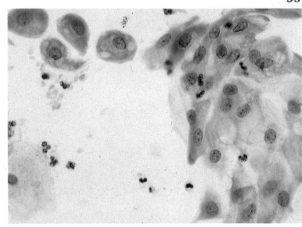

32 Normal smear: appearance of metaplastic cells. (× *160*)

33 Normal smear: appearance of metaplastic cells. (× *125*)

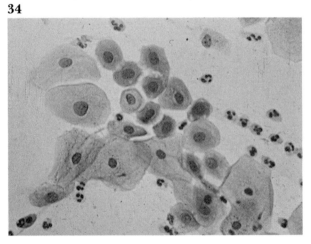

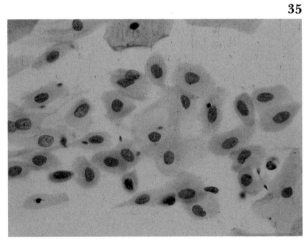

34 Normal smear: appearance of metaplastic cells. (× *160*)

35 Normal smear: appearance of metaplastic cells. (× *160*)

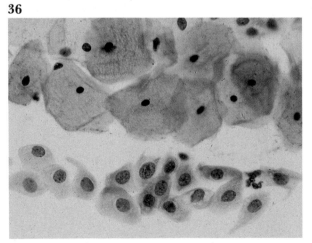

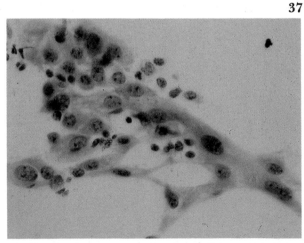

36 Normal smear: appearance of metaplastic cells. (× *160*)

37 Normal smear: appearance of metaplastic cells. (× *160*)

Endometrial cells

It is normal to find endometrial cells in cervicovaginal smears 2-3 days before, during, and for a few days after menstruation. In some women occasional endometrial cells are seen at mid-cycle at the time of ovulation; this may also be associated with mittelschmerz (pain at ovulation). Women wearing an intrauterine contraceptive device (IUCD) commonly shed endometrial cells throughout the cycle, but with this exception endometrial cells found outside the times specified above should be investigated further. In particular, endometrial cells found in postmenopausal smears often indicate disease even when the cells look normal.

38

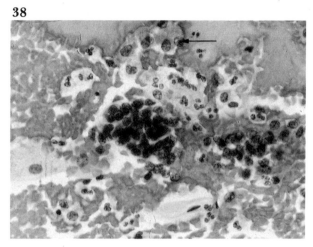

38 Endometrial cells. This field shows a cluster of stromal cells and a few single cells resembling histiocytes (top), which are probably glandular in origin (arrow). (× *160*)

39

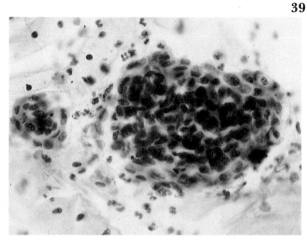

39 Endometrial cells. Clusters of stromal cells are seen, with a suggestion of an acinar form at the periphery. (× *160*)

40

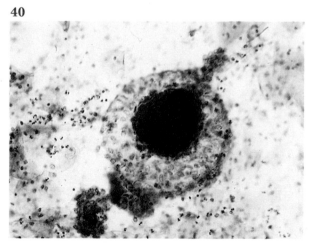

40 Glandular fragment. In this field a glandular fragment has been pulled out of the stroma and presents en fosse with overlayering of the cells at the centre. (× *62*)

41

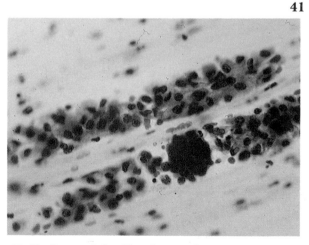

41 Endometrial cells: the exodus. Following menstruation, endometrial cells become histiocytic in appearance, and in some cases these cells are seen streaming away from a more usual glandular or stromal cluster as illustrated in this field. (× *160*)

Normal cells at very high magnification

The next group of pictures demonstrates normal cells at very high magnification (oil immersion lens), to illustrate the difference in nuclear chromatin pattern and cytoplasmic differentiation between cell types.

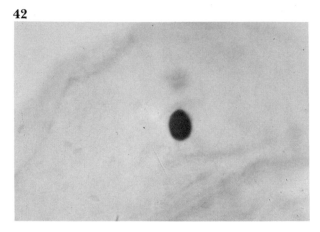

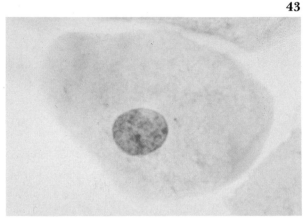

42 Superficial squamous cell. The cytoplasm is translucent, and in this cell it extends beyond the frame of the picture. The nucleus is densely pyknotic and shows no texture; it should be less than 6 μm in diameter. (× 620)

43 Small intermediate cell. A small intermediate cell has been taken to include the whole cell in the frame. Because of this, the cytoplasm is more dense than it would be in a large intermediate cell. The nuclear chromatin pattern is vesicular with occasional chromocentres. A sex chromatin body (Barr body) is seen at the nuclear membrane, reflecting the second X-chromosome of the female. (× 620)

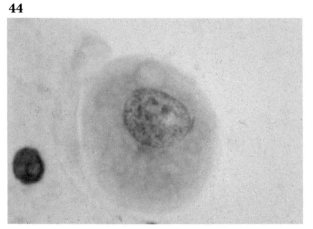

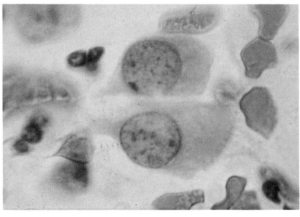

44 Parabasal or metaplastic cell. In this field part of a superficial cell is seen, therefore it seems probable that this is an immature metaplastic cell. Although there is a higher nucleocytoplasmic ratio than in the mature cell, the nucleus still occupies less than half of the cytoplasm. The latter is more dense than it would be in an intermediate cell, and vacuoles are present; this is a common finding, and is probably due to degeneration. The nuclear chromatin pattern is similar to that seen in an intermediate cell. (× 620)

45 Endocervical columnar cell. Two endocervical columnar cells are seen, showing characteristically rounded basal nuclei. The nuclear chromatin pattern is fine but has a more granular distribution than is seen in squamous cells. In addition, there is condensation of chromatin at the nuclear membrane, as well as a sex chromatin body in one cell. The cytoplasm is soft and flocculent; one of the cells shows an end-plate and cilia. (× 620)

46 Endometrial stromal cells. In a cervicovaginal smear endometrial cells have been shed in the menstrual flow, and are more degenerate than when collected by endometrial aspiration. Degeneration causes coagulative necrosis of nuclear chromatin which exaggerates the coarse granularity of the pattern. It will also be noted that these are bare nuclei. (× 620)

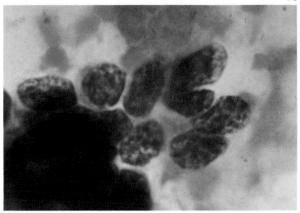

The unsatisfactory smear

For reliable reporting it is essential that the material on the slide is evenly spread, well fixed and not obscured by blood, inflammatory cells or mucus. It should also contain adequate numbers of squamous cells and an endocervical component. The following four views illustrate slides which should be rejected.

47

47 Smear too thickly spread. Some areas are out of focus because of the thickness of the spread. (× 40)

48

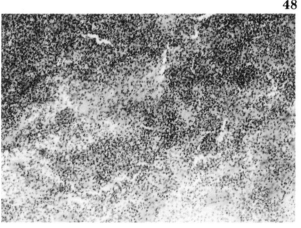

48 Smear obscured by polymorphs. (× 40)

49

49 Smear obscured by blood and inflammatory cells. (× 40)

50

50 Smear obscured by blood and inflammatory cells. This smear has also been allowed to dry before fixation. (× 40)

2 Infection and reactive changes

Reactive changes in cells are usually non-specific and independent of the causative factor. Infection is a common cause, but reaction also follows trauma and concomitant tissue repair. In the case of the cervix, trauma includes therapeutic measures such as cautery, diathermy and laser. Specific cell changes are also seen which, in some cases, will indicate the probable stimulating factor. Such changes are seen after radiotherapy or when cytotoxic drugs or steroids have been given. Virus infections can also produce specific changes in cells. In addition to recognizing reactive changes, it is often possible to identify the infective agent in cervical smears.

Infective agents

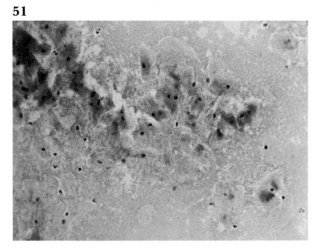

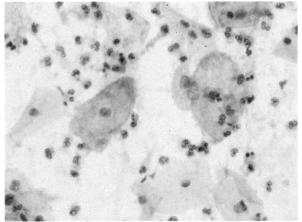

51 Bacterial infection. The normal bacterial flora consists of lactobacilli and these are seen in **6**. In this field there is a heavy coccal haze which indicates bacterial infection, but culture is needed to identify the causative organism. (× *33*)

52 *Gardnerella* infection: clue cells. This organism was originally classified as *Haemophilus vaginalis* and was recognized as causing a specific variety of vaginitis by *Gardner & Dukes (1955)*. It was later reclassified as *Corynebacterium vaginalis* and is now referred to as *Gardnerella* as a tribute to the work of Gardner. In their work Gardner & Dukes described the isolated cells scattered throughout the smear, which have a superimposed layer of coccobacilli. These were called 'clue cells' and, when seen in a smear, indicate the presence of *Gardnerella* infection. However, it is wise to confirm by culture, as work by *Levison et al. (1979)* suggests that the association is not as reliable as was indicated by earlier authors. (× *160*)

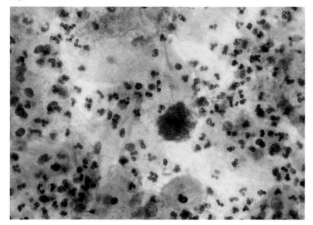

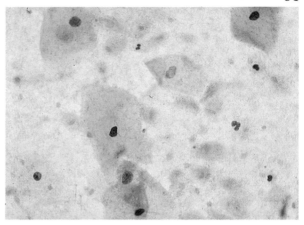

53 Bacterial clusters. Bacteria may present as tight clusters, but this is a non-specific finding and culture is needed to identify the causative organism. (× *160*)

54 *Trichomonas vaginalis.* Trichomonads are a common cause of offensive vaginal discharge and are usually associated with a mixed bacterial infection, as seen in this field. The staining reaction, shape and size of organisms varies, but in this example they are seen as round to pear-shaped with cyanophilic cytoplasm and an eccentric, elongated, small, pale nucleus often referred to as the 'mongol's eye'. Flagellae are not seen in alcohol-fixed material. (× *160*)

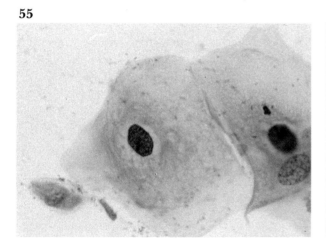

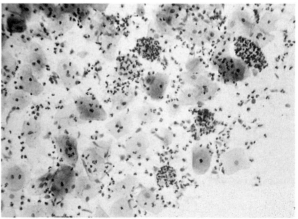

55 *Trichomonas vaginalis.* At higher magnification, this field shows a single trichomonad with intermediate cells which show a narrow perinuclear halo with blurred edges. This is a non-specific inflammatory reaction and must be distinguished from the larger sharp-edged haloes seen in koilocytes (see **127** *et seq.*). (× *400*)

56 *Trichomonas vaginalis*: inflammatory cell pattern. A trichomonal infestation is often associated with an inflammatory cell pattern in which polymorphs surround cells to give a 'buckshot' or 'cannonball' appearance. This has been described as specific for trichomonal infection, but a similar appearance is seen in post-radiation smears (**113**) and it might be expected to occur in some bacterial infections. (× *62*)

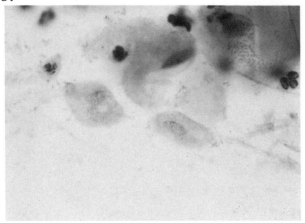

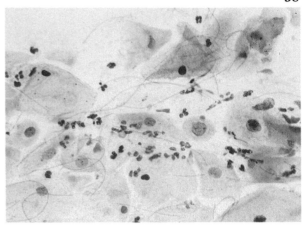

57 _Trichomonas vaginalis._ Occasionally trichomonads have red granules in the cytoplasm, as seen in this field. The significance of this is not known. (× _400_)

58 _Trichomonas vaginalis_ and _Leptothrix._ Trichomonads are often seen with saprophytic _Leptothrix._ In this field the _Leptothrix_ is seen as branching threads which present as a single line. They must be distinguished from _Candida_ (see **60** _et seq._) (× _160_)

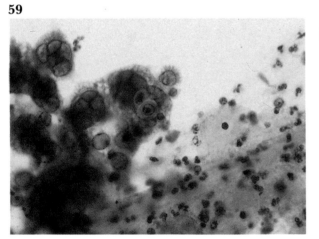

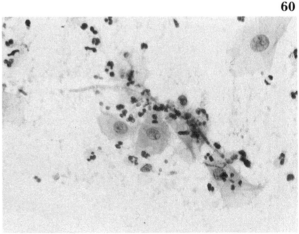

59 _Trichomonas vaginalis_ and herpes. Multiple infections are common. In this field trichomonads, with an eosinophilic cytoplasm, are seen together with a bacterial haze and cells showing the cytopathic effect caused by herpes virus infection (see **121** _et seq._). (× _160_)

60 _Candida._ This is a fungal infection which is common during pregnancy and in women taking oral contraceptives. It causes thick, white, cheesy discharge and is colloquially referred to as 'thrush'. In cervical smears it is preferable to identify both spores and hyphae, as seen in this field, before making the diagnosis. The hyphae are seen to be segmented and appear as a double 'tram line'. Compare with _Leptothrix_ in **58**. (× _160_)

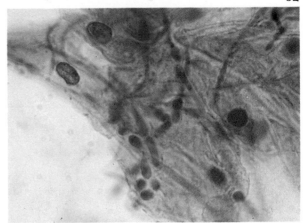

61 *Candida.* In addition to spores and hyphae, this field shows clumping of squamous cells. This can alert the screener to search for the presence of *Candida*. (× *160*)

62 *Candida.* Higher magnification shows the morphology of spores and hyphae more clearly. For precise identification of fungal type, culture is needed. The *Candida* group is found most commonly, so some laboratories use this terminology. Others prefer to be less specific and report 'fungal spores and hyphae seen'. (× *400*)

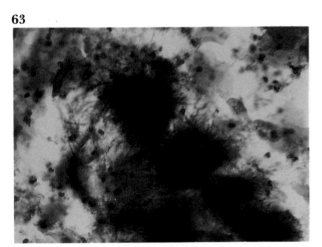

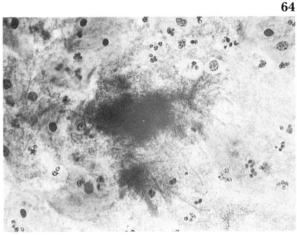

63 *Actinomyces.* *Actinomyces* are found in smears taken from women wearing an IUCD (*Gupta et al., 1976*). When associated with pelvic pain and irregular bleeding it may be the cause of pelvic inflammatory disease. In these cases the IUCD is removed for confirmatory culture, but in the absence of symptoms it is usual to note the finding without further action. (× *400*)

64 *Actinomyces.* Another example, with blurred eosinophilic staining. (× *160*)

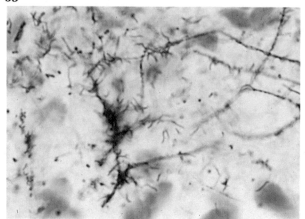

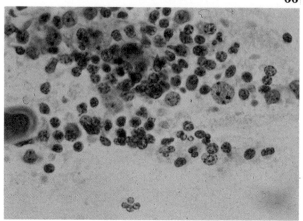

65 *Actinomyces*. Confirmation by culture is difficult unless the IUCD can be removed and cultured. The morphological appearance of peripheral threads extending from a dense central area can be mimicked by collections of debris. However, these are usually Gram-negative, whereas *Actinomyces* is Gram-positive. This Gram-stained smear shows a positive reaction. In this case the presence of *Actinomyces* was confirmed by culture. (*Gram stain*, × 620)

66 *Chlamydia*. A firm diagnosis of *Chlamydia* must be by culture (which is difficult) or by antigen/antibody response. In cervical smears morphological features which cause suspicion of *Chlamydia* include a lymphocytic reaction similar to that of follicular cervicitis. This is the presentation seen in this field. In addition, free eosinophilic coccoid elementary bodies are seen. It is more specific when these bodies are seen in the finely vacuolated cytoplasm of infected cells, and more particularly when moulded perinuclear vacuoles which contain inclusion bodies are present. This feature is not seen in this example, but the presence of *Chlamydia* was confirmed by immunofluorescence (*Gupta et al., 1979*). (× 250)

Non-specific reactive and degenerative changes

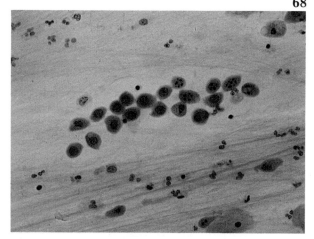

67 Mitoses. Mitoses are less commonly seen in cytological material than in tissue, but they can be found when there is rapid tissue regeneration. It is more usual to find them in tissue fragments, but this example is in a single parabasal cell. In benign reaction mitoses are usually normal. (× 400)

68 Karyorrhexis. It is necessary to distinguish mitoses from degenerative fragmentation of the nucleus as seen in the parabasal cells in this field. This form of degeneration is described as karyorrhexis; other forms include karyolysis (**69**) and cytolysis (**6**). (× 120)

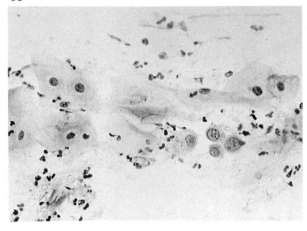

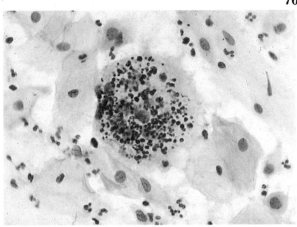

69 Karyolysis. The nucleus of the cell at the centre of the field has almost disappeared. Its position is identified by a pale, incomplete nuclear membrane. (× *125*)

70 Phagocytosis. Giant multinucleated histiocytes are seen in smears from patients with chronic infection or a granulomatous reaction. These may phagocytose nuclear debris, as seen in this field. (× *160*)

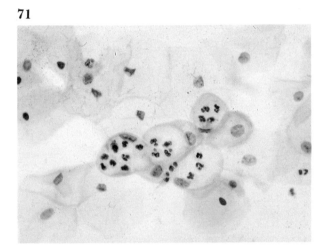

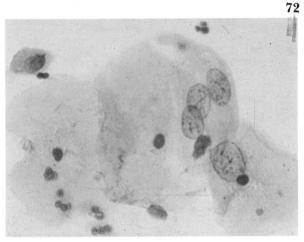

71 Cytoplasmic vacuoles containing polymorphs. An appearance resembling phagocytosis is seen when polymorphs invade degenerative vacuoles in squamous or columnar cells. The cells in the vacuoles are sometimes referred to as engulfed cells. Note that in this field the invading polymorphs look viable when compared with the nuclear fragments in **70**. (× *160*)

72 Multinucleation. When there is rapid tissue regeneration nuclei may replicate more quickly than the cytoplasm divides. This results in multinucleation with benign-looking nuclei which have identical appearances. In this field there is a trinucleate cell with enlarged hypochromatic nuclei, and below this is another cell with a single enlarged hypochromatic nucleus. Similar changes have been reported with folic acid deficiency (*van Niekerk, 1966*). (× *250*)

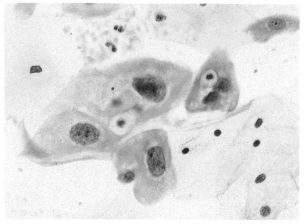

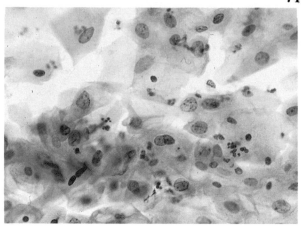

73 Reactive hyperplasia. Enlarged nuclei are common in cells showing reactive changes. The nuclear chromatin may be dispersed so that the nucleus appears hypochromatic. In other cases degeneration can cause coagulative necrosis, and the nuclear chromatin becomes hyperchromatic and blurred. Both nuclear appearances are seen in this field. In addition, perinuclear haloes are seen, together with debris contained in vacuoles in the cytoplasm. One cell is binucleate. (× *160*)

74 Reactive changes: squamous cells. In this field nuclei are large, with dispersion of chromatin giving a granular appearance. Some nuclear degeneration is present and cytoplasmic staining is blurred. (× *160*)

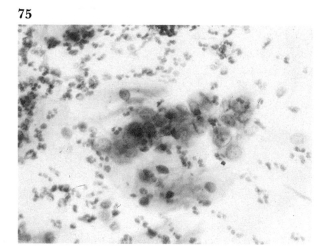

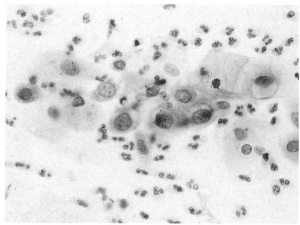

75 Reactive changes: endocervical columnar cells. In this field a strip of endocervical columnar cells shows enlarged, pale nuclei with prominent nucleoli. The staining is generally blurred and in some cells the cytoplasmic staining is amphophilic. Polymorphs are present and many are degenerate. Poorly stained, blurred smears are commonly seen with infection and reports may be unreliable. Smears should be repeated after treatment of the infection. (× *160*)

76 Reactive changes: metaplastic cells. In this field metaplastic cells have large, granular nuclei. Perinuclear haloes are seen in a few cells, and there is some cytoplasmic vacuolation. Comparison with **156-159** will show that when there is increased granularity of nuclear chromatin the changes begin to border on dyskaryosis. (× *160*)

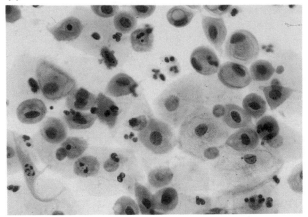

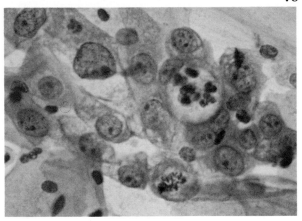

77 Reactive changes: metaplastic cells. In this field there is more nuclear degeneration and hyperchromasia; karyorrhexis is seen in occasional cells. The cytoplasm is dense and amphophilic in some cells. (× *160*)

78 Reactive changes: endocervical columnar cells. At higher magnification this cluster of endocervical columnar cells shows marked reactive changes, with prominent nucleoli which vary in size and number from cell to cell. A polymorph-containing vacuole is present and a normal mitosis. This degree of reactivity can cause difficulties in diagnosis, but the presence of regular, finely granular chromatin should be noted as evidence that these cells are benign. (× *250*)

Surface reaction

This can result from persistent trauma and chronic infection, and in particular from human papillomavirus (HPV) infection. The histological picture is of hyperkeratosis, with anucleate keratinized squamous cells, or parakeratosis when keratinized cells with degenerate pyknotic nuclei are seen. Surface reaction of this type can be non-specific and benign, or associated with HPV infection, but in both cases the thickness of the protective layers can prevent exfoliation of underlying abnormal cells. Persistence of this finding makes further investigation by colposcopy and biopsy advisable.

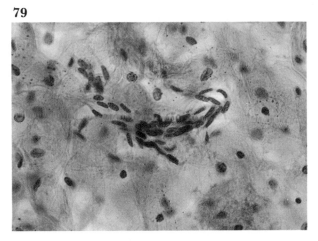

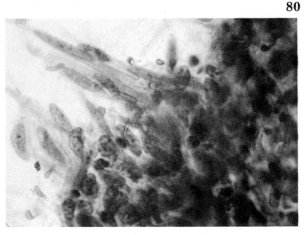

79 Fibre cells. This field shows elongated fibre cells, which is one way in which surface reaction can present in a cervical smear. (× *160*)

80 Infected cervical polyp. At higher magnification similar fibre cells, showing prominent nucleoli, are seen at the surface of a tight cell cluster. This patient had a benign infected cervical polyp with foci of granulation tissue. (× *250*)

81

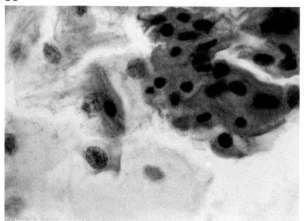

81 Dyskeratotic cells (parakeratosis). This field shows a sheet of keratinized cells with large, degenerate, pyknotic nuclei. Squamous cells showing pale, enlarged nuclei are also present. This picture could reflect HPV infection, but in this case it was the result of trauma to a large cervical polyp presenting at the introitus. (× *250*)

82

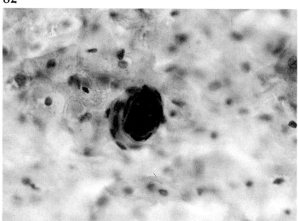

82 Epithelial pearl. Epithelial pearls are usually benign and result from scraping the surface of a crypt opening which has been blocked by squamous metaplasia. They can also reflect simple epithelial knots in original squamous epithelium. In these cases the nuclear chromatin looks benign, as in this field. (× *160*)

83

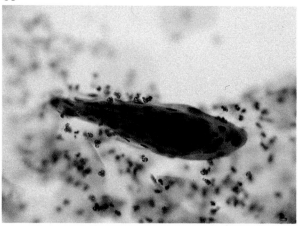

83 Dyskeratotic epithelial pearl. This picture is introduced for comparison with **82**. It can be seen that the cytoplasm is keratinized, and the nuclei are hyperchromatic and pleomorphic. This appearance is more likely to be associated with underlying disease such as warty keratosis. (× *120*)

Illustrative cases

Reaction to IUCD

Some women who wear an IUCD show marked reactive changes, both degenerative and regenerative, in metaplastic cells and endocervical columnar cells. In some cases these changes can be so marked that the appearances cause suspicion of dyskaryosis, but they regress when the IUCD is removed. Endometrial reaction also occurs because the IUCD acts as a foreign body, causing a minor degree of endometritis. Endometrial cells may exfoliate throughout the cycle, and when severe reactive changes are present the appearances can cause suspicion of endometrial hyperplasia or even neoplasia. Reaction can be so marked that psammoma bodies form and exfoliate in the smear (*Highman, 1971*).

84

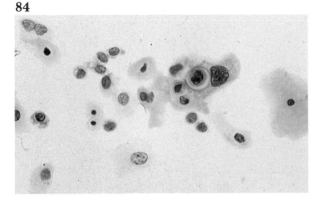

84 IUCD reaction: metaplastic cells. This field shows marked hyperplasia of metaplastic cells, with some nuclear degeneration. (× *160*)

85

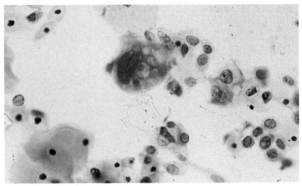

85 IUCD reaction: endocervical columnar cells. Elsewhere in the same smear reactive endocervical columnar cells were found, together with a multinucleated cell. It is possible that constant minor trauma from the IUCD thread is the cause of these changes. In this case the smear became negative 3 months after the IUCD was removed. (× *160*)

86

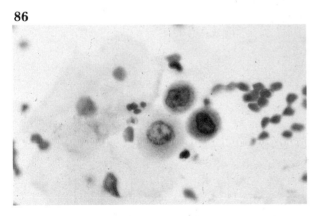

86 IUCD reaction: endometrial cells. Single atypical endometrial cells are seen in this field. Note the irregularity of nuclear membrane and prominent nucleoli. These changes caused suspicion of an endometrial lesion and dilatation and curettage was thought to be necessary. (× *250*)

87

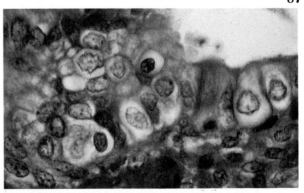

87 Endometrial curettings from the same case. At the same magnification the surface of the endometrium shows reactive hyperplasia with cells of the type seen in the smear. This was the only abnormality found and was reported to be caused by the IUCD. Further follow-up was negative. (*H&E,* × *250*)

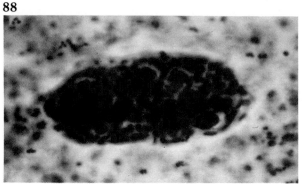

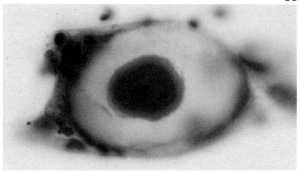

88 IUCD reaction: psammoma bodies. In this field multiple psammoma bodies form a tissue fragment. The cells comprising the fragment are otherwise regular and look benign. (× *160*)

89 IUCD reaction: psammoma body. At a higher magnification a single psammoma body is seen in another field from the same smear. Further investigation showed that the patient was negative for endometrial or ovarian carcinoma. (× *400*)

Atrophic vaginitis

Smears from postmenopausal women with atrophic vaginitis show a wide range of cellular changes resulting from associated non-specific infection. In some cases the degree of abnormality can be so marked as to cause suspicion of dyskaryosis. With this degree of abnormality it is advisable to repeat the smear after the use of oestrogen vaginal cream for 4 weeks. The treatment causes maturation of the squamous epithelium by the direct effect of oestrogen on the target organ (vagina). This is more reliable than systemic oestrogen, as the response of the target organ varies with systemic treatment. In addition, atrophic epithelium is thin and easily traumatized. The damage may be microscopic, but is reflected by the content of the smear. In such cases histiocytes and polymorphs are present with multinucleated histiocytes and fibroblasts, which suggests a granulomatous reaction as seen in granulation tissue. Clinical ulceration also results in a characteristic cell picture, and this can occur in women with ulcerated procidentia or who are wearing a ring pessary.

90

91

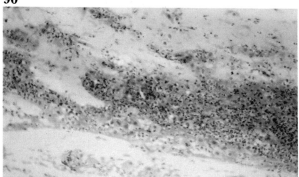

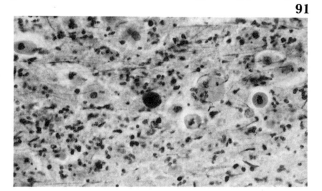

90 Atrophic vaginitis. This field shows a common presentation with a serous background and red cells and inflammatory cells obscuring an atrophic smear. The staining reaction is poor because of some drying before fixation. (× *40*)

91 Atrophic vaginitis: 'blue blobs'. This is another example of atrophic vaginitis with inflammatory cells, much background debris and parabasal cells showing a variety of degenerative changes. At the centre of the field is a round structure staining blue-black. These are found in atrophic smears and have caused suspicion of malignancy. The cause is uncertain, although various theories have been put forward (*Ziabkowski & Naylor, 1976*). It is most likely that this is a form of DNA or RNA breakdown rather than the presence of inspissated mucus. (× *160*)

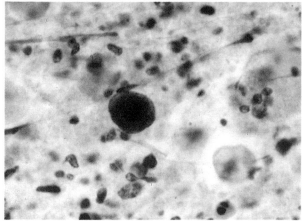

92 Atrophic vaginitis: 'blue blobs'. At higher magnification the appearances suggest that this is the outline of a cell obscured by dense, finely powdered material which stains dark blue. (× *400*)

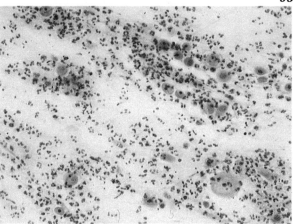

93 Atrophic vaginitis. Another example, showing a poor staining reaction. The background consists mainly of polymorphs and nuclear debris. Some parabasal cells show large, pale, reactive nuclei, whereas in others the nuclei have become pyknotic. In spite of the poor staining reaction, keratinization of the cytoplasm can be recognized in some cells. (× *80*)

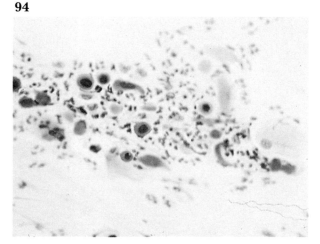

94 Atrophic vaginitis: borderline changes. This field demonstrates the degree of abnormality which can be seen in benign atrophic smears. Karyolysis and nuclear degeneration are present and bizarre cell shapes are seen. In some cells the coagulative necrosis causes nuclear irregularity which is borderline for dyskaryosis. (× *125*)

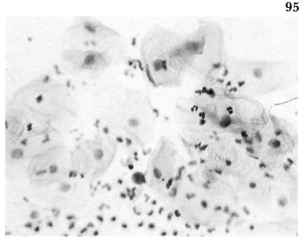

95 The same case after oestrogen vaginal cream. The picture in this field is quite different. There is less inflammatory cell exudate and most of the cells present are normal intermediate cells. One keratinized parabasal cell remains. If dyskaryotic cells had been present in this case, they would now be apparent against a background of normal mature cells. Oestrogen vaginal cream was used for 4 weeks in this case. (× *125*)

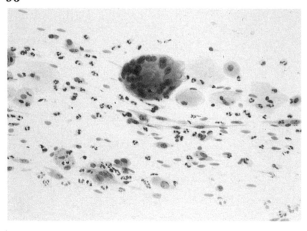

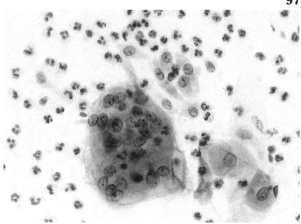

96 Atrophic vaginitis with granulation tissue.
In this example the few squamous cells present show a range of maturation. Histiocytes and polymorphs are present, with a giant multinuclear histiocyte and elongated cells with soft cytoplasm, which are fibroblasts. This picture suggests a granulomatous reaction, which in this situation is most likely to be non-specific granulation tissue. (× *80*)

97 Atrophic vaginitis with granulation tissue.
This field shows a similar picture, with a giant multinucleated cell overlying squamous cells. (× *160*)

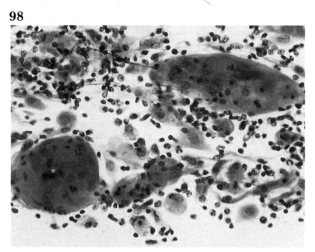

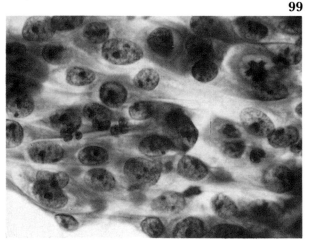

98 Vaginal ulceration. A more florid picture is seen when frank ulcers are present; this woman wore a ring pessary to control prolapse. The smear contains many multinucleated histiocytes with fibroblasts, histiocytes and polymorphs. (× *125*)

99 Ulcerated procidentia: epithelial regeneration. In cases with clinical ulceration it is common to have tissue regeneration at the edge of the ulcer. When these cells are seen in the smear they can cause suspicion of malignancy. This example is typical of such a sheet of cells which presents as a monolayer of pavement cells. The appearance is reminiscent of the edge of a growing tissue culture. There is some pleomorphism, and nucleoli are prominent. In some cells these are large and in other cells they are multiple; a mitosis is also present. The main feature in favour of a benign lesion is the pale, regular, dispersed chromatin pattern of the nuclei. It should also be noted that the nucleoli are regular in shape and the mitosis is normal. (× *250*)

Follicular cervicitis

This is a lymphocytic reaction which is usually seen deep to cervical epithelium, but has been reported deep to vaginal epithelium. The collections of lymphocytes present as normal follicles with mature lymphocytes at the periphery and immature forms at the centre.

The condition is more common in postmenopausal women, but is also seen as a reaction to *Chlamydia* infection (**66**). In these cases the cervix is granular on colposcopic examination.

100

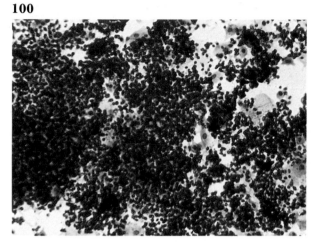

100 Follicular cervicitis. The cytological picture seen in this field is characteristic, with sheets of small hyperchromatic cells which have a very high nucleocytoplasmic ratio. In this case the cells present are mainly mature lymphocytes, but when there is a preponderance of immature cells this can cause suspicion of lymphoma. (× *62*)

101

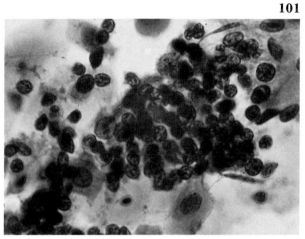

101 Follicular cervicitis. At higher magnification another field from the same case shows immature forms. (× *250*)

102

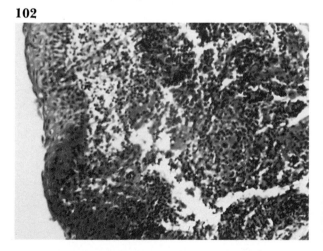

102 Follicular cervicitis: colposcopic biopsy. On colposcopic examination an area of vascular granularity was seen on the cervix and this was biopsied. The section shows traumatized tissue, but the remains of a lymph follicle deep to red, necrosed surface epithelium can be recognized. (*H&E*, × *62*)

103

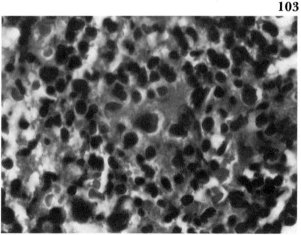

103 Follicular cervicitis: colposcopic biopsy. This field shows the centre of the follicle at the same magnification as **101**, which permits comparison of the cells. (*H&E*, × *250*)

Decidual reaction

During pregnancy foci of decidua can form in any epithelium which derives from the Müllerian duct. When this occurs on the cervix the cells scraped from the area are characteristic but can cause concern if not recognized.

104

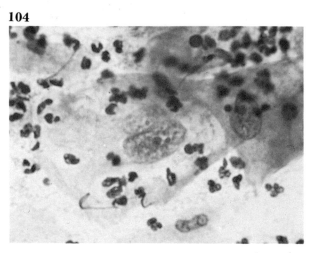

104 Decidual reaction. This field shows decidual cells. One is partly obscured by a squamous cell, but the other can be seen to be a large cell with clear cytoplasm and a pale, granular nucleus with large, pink nucleoli. In this case the nucleus is binucleate. (× 250)

105

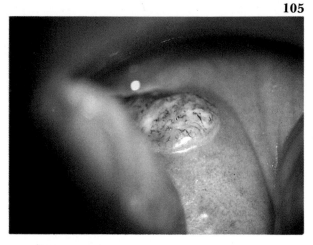

105 Decidual reaction: colposcopy. In this case the decidual reaction presented as a white, dome-shaped lesion on the cervix which could be seen with the naked eye and caused clinical suspicion of neoplasm. (*Saline, × 16*)

106

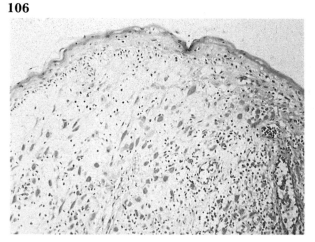

106 Decidual reaction: colposcopic biopsy. Directed biopsy shows thinning of squamous epithelium over oedematous stroma which contains decidual cells. (*H&E, × 62*)

107

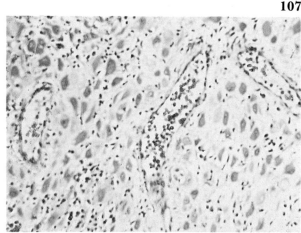

107 Decidual reaction: colposcopic biopsy. At higher magnification the decidual cells in the biopsy can be compared with the cells in **104**. (*H&E, × 125*)

Vaginal adenosis

Vaginal adenosis can occur as a congenital abnormality and also with exposure of the fetus to diethylstilboestrol (DES) in utero (*Prins et al., 1976*). Because of the association of DES exposure with adenocarcinoma of the vagina, girls known to have been exposed in this way have been monitored by vaginal aspiration smears. The presence of columnar cells alerts to the need for further investigation. In other cases, some non-DES exposed, vaginal examination for other reasons has shown abnormality in the fornices. The case illustrated here fell into this category; she had not been exposed to DES.

108

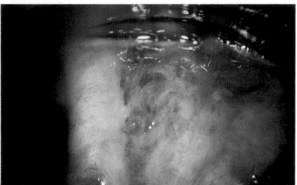

108 Vaginal adenosis: colposcopy of posterior fornix. The posterior fornix looks vascular and translucent. (*Saline, green filter, × 16*)

109

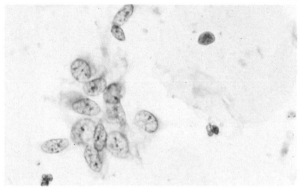

109 Directed smear from the posterior fornix. This smear contains cells resembling endocervical columnar cells which would not be expected in a directed smear from the posterior fornix. (*× 250*)

110

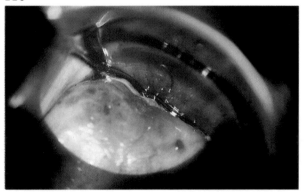

110 Healing vaginal adenosis: colposcopy. Six months later it can be seen that most of the columnar epithelium is replaced by squamous metaplasia. (*Saline, green filter, × 10*)

111

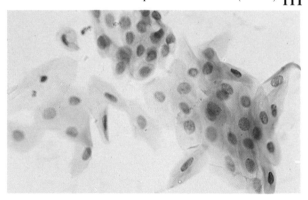

111 Directed smear from posterior fornix. The colposcopically directed smear is now found to contain metaplastic cells. (*× 250*)

112

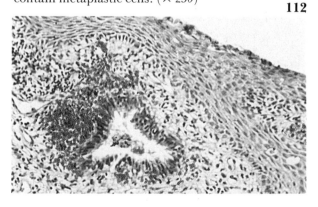

112 Colposcopic biopsy of posterior fornix. This section is from the colposcopic biopsy taken at the second visit. The surface is covered by intact squamous epithelium, and a residual crypt is seen in the stroma. (*H&E, × 62*)

Examples of iatrogenic cytology

The cell changes shown in this section are caused by therapy. It will be noted that there are similarities in the appearance of cells caused by different forms of treatment, as in all cases basic disturbance of cell metabolism will have occurred. Common features are very large cells (macrocytosis) and bizarre cell forms. Similar cell changes are seen in women with folic acid deficiency where there is disturbance of DNA metabolism (*van Niekerk, 1966*) and a common aetiology is probable.

113

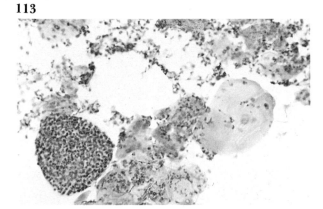

113 Radiation effect. Bizarre cell changes are common for many years following radiotherapy. This field contains macrocytes, and one cell is covered by closely packed degenerate polymorphs (compare with **56**). (× *62*)

114

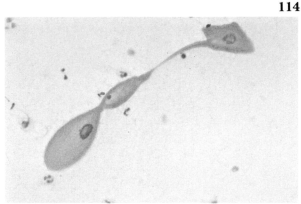

114 Radiation effect. This field shows a bizarre cell form, probably caused by disturbance of normal cell division. (× *62*)

115

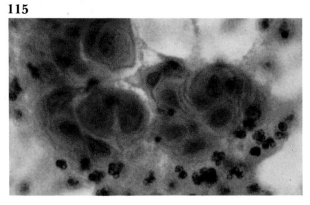

115 Radiation effect versus malignancy. Cell changes as illustrated in **113** and **114** do not present problems in diagnosis, but when follow-up smears are taken soon after radiotherapy it can be difficult to distinguish recurrent tumour from radiation effect. This field shows acinar-like clusters in a woman treated for squamous cell carcinoma of the cervix 6 months previously. Clinically the findings caused suspicion of recurrence, but the cytological picture could be said to be more suggestive of an adenocarcinoma. However, the presence of concentric rings in the cytoplasm is really a squamous cell phenomenon, and this was confirmed by the biopsy which showed radiation changes in granulation tissue with no evidence of residual or recurrent neoplasm. (× *250*)

116

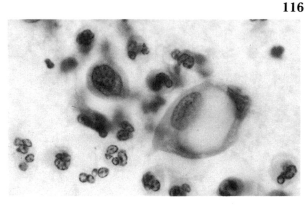

116 Malignant cell plus radiation effect. This picture is included for comparison. It is again a question of recurrent squamous cell carcinoma of cervix following radiotherapy. In this field a malignant squamous cell is seen with a vacuolated macrocyte. (× *400*)

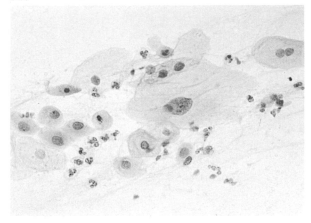

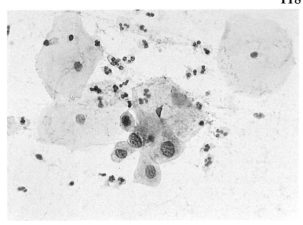

117 Cytotoxic drugs. Note the similar changes seen in this smear taken from a woman who was receiving treatment for leukaemia. (× *125*)

118 Cytotoxic drugs. This is another example of a cervical smear taken from a woman receiving treatment for leukaemia. In this field the affected cells are of the parabasal type. Note the foamy cytoplasm within the sharp cytoplasmic border and the granular nuclei. (× *125*)

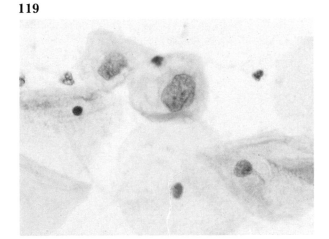

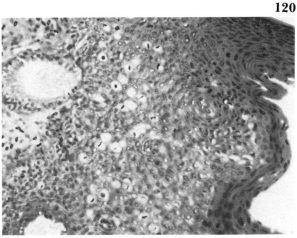

119 Effect of silver nitrate cauterization of cervix. Macrocytosis and bizarre cell forms are also seen after tissue damage by cautery, diathermy and laser. This example is seen in a cervical smear taken from a woman 6 months after silver nitrate cauterization of the cervix. In this field there is macrocytosis of parabasal cells. (× *250*)

120 Colposcopic biopsy. Because of the cytological abnormalities the woman was examined colposcopically, and biopsy shows the very marked reactive hyperplasia of metaplastic epithelium. (*H&E*, × *62*)

3 Cell changes caused by virus infection

Herpesvirus and HPV

The recognition of human papillomavirus and herpesvirus as carcinogens or co-carcinogens in the development of squamous cell carcinoma of the lower female genital tract has made it important to recognize the presence of these infections by the cell changes seen when the viruses are present. In both instances the viral particles invade cell nuclei and cause characteristic degenerative changes as the natural history of the virus proceeds. When nuclear and cell degeneration are complete, viral particles are released to enter other cells and repeat the cycle. Natural immunity can make the acute phase self-limiting but viral DNA may have entered the genome of replicating nuclei, giving a cell line with the potential to transform later to a neoplastic cell line. Secondary 'trigger' mechanisms which have been implicated include parity (*Miller et al., 1980*), radiotherapy (*Choo & Anderson, 1982*), smoking (*Trevathan et al., 1983*) and immuno-suppression (*Stanbridge & Butler, 1983*).

The presence of herpesvirus can be confirmed by culture and rising antibody levels, but human papillomavirus (HPV) has not been cultured. Although the presence of koilocytes (illustrated below) is usually considered to be pathognomonic of HPV infection, confirmation can be obtained by identifying viral particles using transmission electron microscopy (TEM), by identifying antigens using immunocytochemistry, or by using DNA probes to identify the precise serotype.

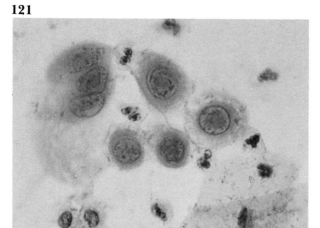

121 Herpesvirus changes. This field shows single cells in which some nuclei are pale, whereas others show nuclear degeneration in which the chromatin seems to be coagulating centrally and also at the nuclear membrane. (× *250*)

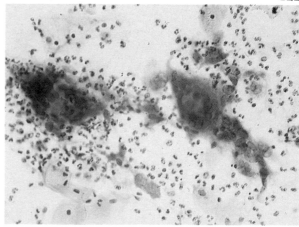

122 Herpesvirus changes. At a later stage multinucleation appears, with close-packed, moulded nuclei. The nuclei are pale with a ground-glass appearance. (× *80*)

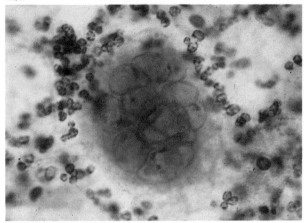

123 Herpesvirus changes. At a higher magnification these changes are seen more clearly in a single multinucleated cell. (× *320*)

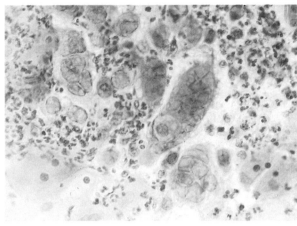

124 Herpesvirus changes. A common feature is blurring of the staining of affected cells with 'folding' of the nuclei. (× *160*)

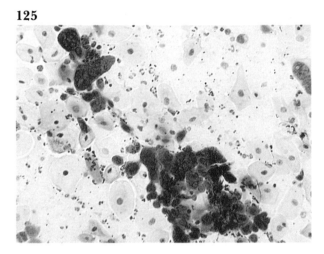

125 Herpesvirus changes (courtesy Mr Reddish, Oldham General Hospital). This smear was collected from a patient with acute vulval herpes. The staining of affected cells is blurred and also amphophilic. (× *62*)

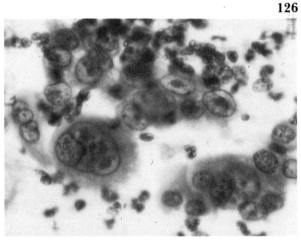

126 Herpesvirus changes: intranuclear inclusions. Herpesvirus, as well as related viruses such as cytomegalovirus, can also present with intranuclear inclusions. This is thought to be a late stage of degenerative change when viral particles have left the cell, and the inclusion consists of residual nuclear material. Note that there is also condensation of chromatin at the nuclear membrane with a clear halo and the inclusion at the centre of the nucleus. This should not be confused with a prominent nucleolus. Multinucleated cells are also present. (× *250*)

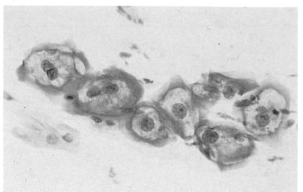

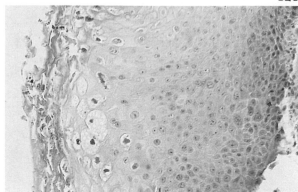

127 Human papillomavirus infection (HPV): koilocytes. *Purola & Savia (1977)* and *Meisels and co-workers (1976, 1977)* recognized koilocytes as diagnostic of HPV infection. It was also shown that the infection could be present in flat epithelium without overt warts. This field shows typical koilocytes with enlarged, reactive nuclei and the characteristic deep halo separated from the cytoplasm by a sharply condensed border. These haloes must be distinguished from the narrow perinuclear haloes with a blurred border seen with many infections as a non-specific reaction (see **55**). (× 125)

128 HPV infection: colposcopic biopsy. This field shows tissue from the case illustrated in **127**. Koilocytes are seen near the surface and the surface layers are degenerate and keratinized. Otherwise the tissue shows reactive changes only. (*H&E, × 62*)

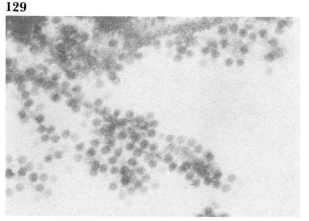

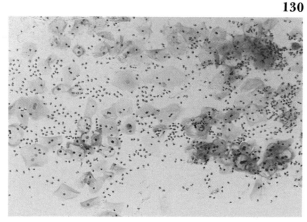

129 Virus particles: TEM. This field shows viral particles in the case illustrated in **127** and **128**. Virus particles are scanty in genital warts and, although confirmation is possible by this method, it is tedious and time-consuming. Identification is made by the morphology and measurements of the particles and confirmed by negative staining (*Stanbridge et al., 1981*). (× 25,000)

130 HPV infection: koilocytes and single keratinized cells. The surface of HPV-infected epithelium commonly shows hyperkeratosis and parakeratosis (see **128**). Consequently, keratinized anucleate squames and dyskeratotic cells are seen in cervical smears, in addition to koilocytes. In some cases hyperkeratosis and parakeratosis are present in a thick layer which prevents exfoliation of koilocytes, so making a firm diagnosis of HPV infection impossible. However, in recent years HPV infection has become so common and its association with intraepithelial neoplasia so close that repeated smears containing keratinized cells warrant referral for colposcopy and biopsy. Koilocytes are present in this field, but note the occasional single keratinized cells. (× 40)

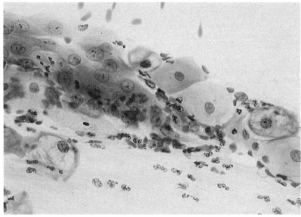

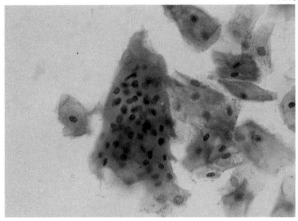

131 HPV infection: koilocytes and non-specific inflammatory changes. Non-specific reactions are seen in HPV infected smears in the same way as when infection is caused by another organism. Note the sheet of cells with enlarged, pale nuclei and prominent nucleoli. (× *120*)

132 HPV infection: dyskeratotic spike. This field shows a syncytium of dyskeratotic cells in the form of a spike. It is from the surface of a wart covered by keratinized spicules. (× *120*)

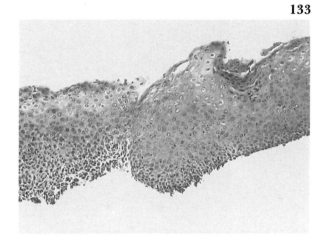

133 HPV infection: colposcopic biopsy. Section of this biopsy shows a spicule which could present in a cervical smear in the same way as the spicule in **132**. (*H&E*, × *37.5*)

HPV infection with borderline changes

Previous illustrations in this series have presented cases in which HPV infection was the only abnormality. Diagnostic problems arise when there is, in addition, nuclear abnormality bordering on dyskaryosis. In cases where diagnostic dyskaryotic cells are seen, management of the case will follow that which is usual for the degree of intraepithelial neoplasia anticipated (*Kaufman et al., 1983*) and, even when changes are borderline, referral for colposcopy and biopsy is advisable as surface keratosis may prevent exfoliation of the most abnormal cells.

134

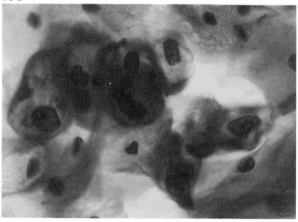

134 HPV infection: mild dyskaryosis. In this case the cervical smear contained koilocytes with mildly dyskaryotic cells. (× *320*)

135

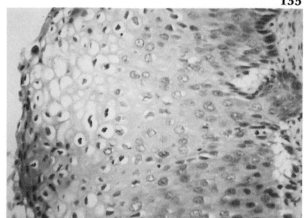

135 HPV infection with CIN I: colposcopic biopsy. This is the biopsy taken from the case illustrated in **134**. CIN I is seen in infected tissue. (*H&E*, × *80*)

136

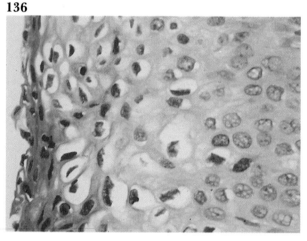

136 HPV infection with CIN I: colposcopic biopsy. At higher magnification the marked nuclear abnormality present in the koilocytes can be seen. (*H&E*, × *320*)

137

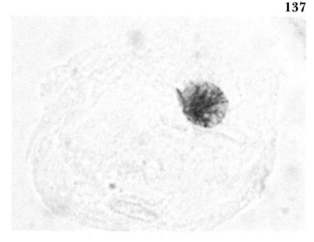

137 HPV infection. For completeness, an example of immunocytochemistry in the identification of HPV infection is shown. This field shows a positive reaction in a single squame present in a cervical smear. (*Immunoalkaline phosphatase*, × *400*)

138

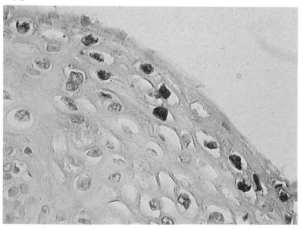

138 HPV infection: biopsy of cervix. A similar reaction is seen in a section from the cervical biopsy. (*Immunoalkaline phosphatase*, × *160*)

4 Contaminants

Unexpected organisms and structures can be found in cervicovaginal material. In some cases this is caused by spread of extragenital infection or contamination from vulval or perianal skin, and in others it is caused by contamination of glass slides before use or during processing.

139

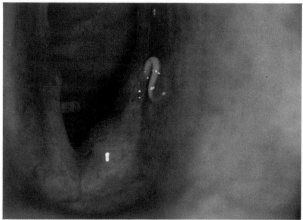

139 Oxyuris (threadworm): colposcopic picture. Oxyuris ova can be found as a contaminant in cervicovaginal smears taken from women or girls with an intestinal infection. In the case illustrated Oxyuris had also caused a vulvovaginitis in a child aged 7 years. The heat from the colposcope lamp caused the worm to present at the introitus. (*Saline*, × 25)

140

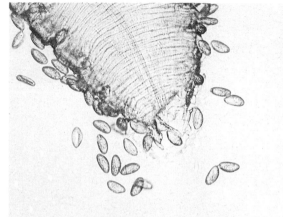

140 Oxyuris (threadworm). This picture illustrates the worm seen in **139**, which has been mounted for photography. Compression has caused the ova to spill from the body of the worm so that the size of the ova can be evaluated. (*Unstained*, × 40)

141

141 Oxyuris ova. The more usual presentation of ova is in a Papanicolaou stained smear when the internal structure, surrounded by a chitinous capsule, can be seen. It is important to note the large size of these ova compared with normal squamous cells. Similar but smaller bodies are sometimes found in smears and mistaken for Oxyuris ova. (× 160)

142

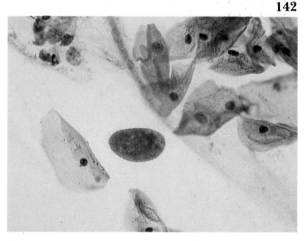

142 *Balantidium coli.* This is an example of a structure which could be mistaken in this way. The error should not occur if the smaller size is noted, together with the absence of both the chitinous capsule and detailed internal structure. (× 160)

143

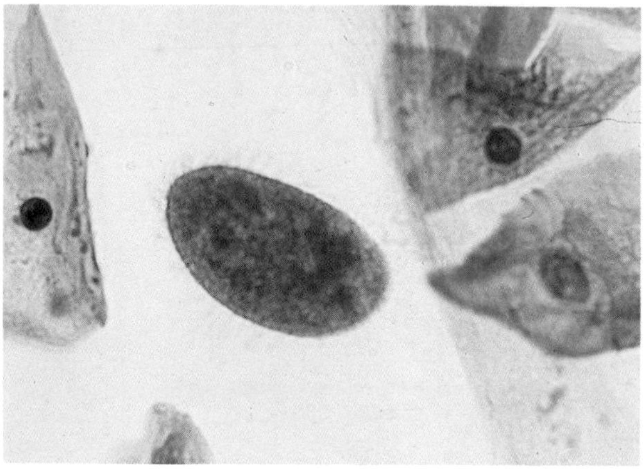

144

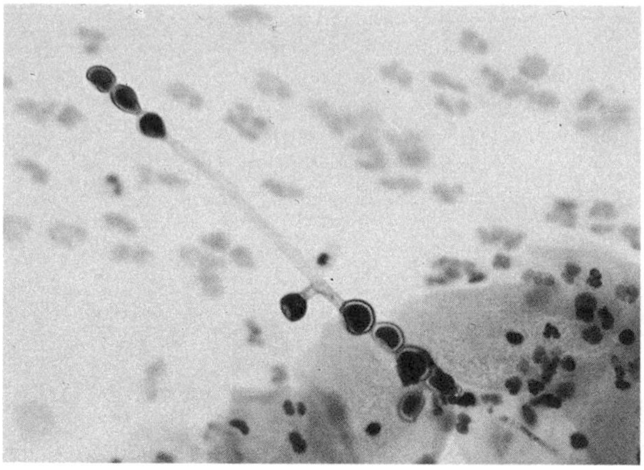

143 ***Balantidium coli.*** In addition, at higher magnification a ciliated border is seen. (× *320*)

144 Aspergillosis sporangium. This is an example of a fungal contaminant. (× *160*)

145

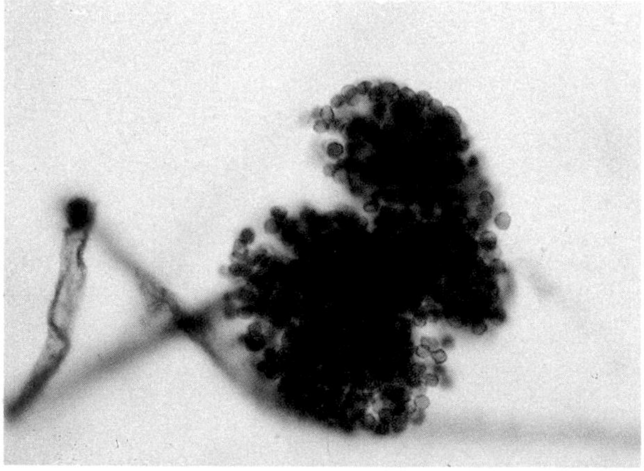

146

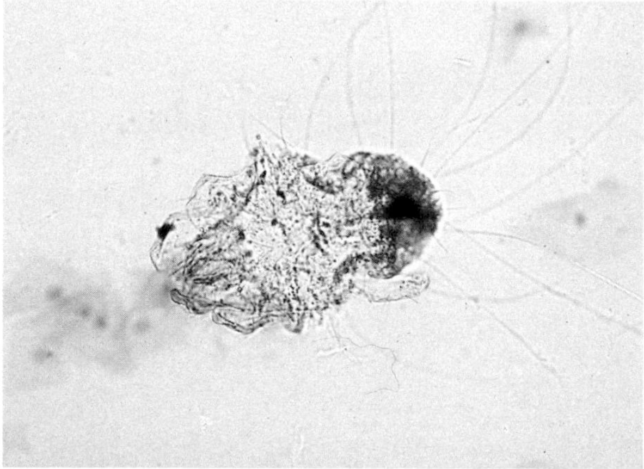

145 Aspergillosis sporangium. At higher magnification the breakdown of one of the heads, with the release of spores, is seen. (× *400*)

146 Crab louse. (× *80*)

147

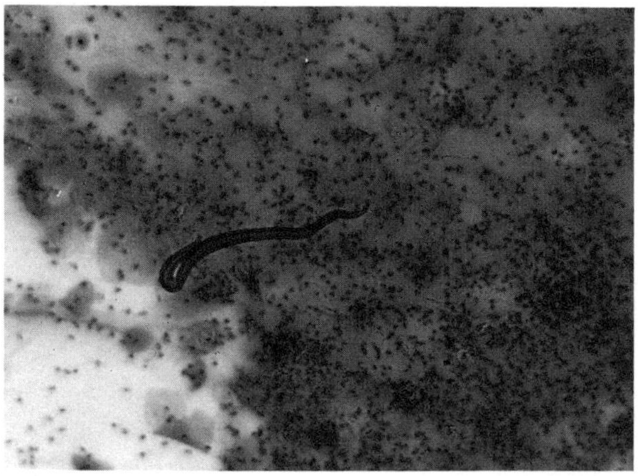

148

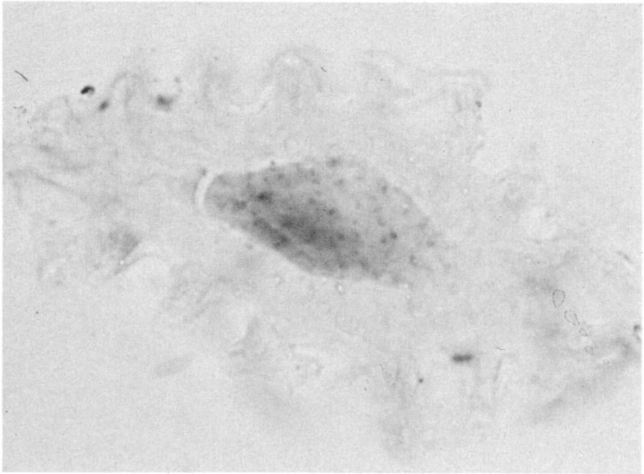

147 Strongyloides. (× *33*)

148 Plant stone or sclerid. (× *250*)

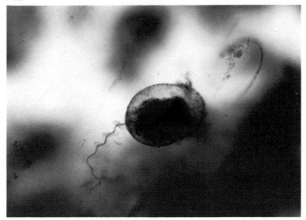

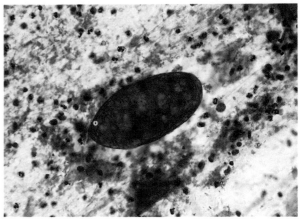

149 Vorticella. This is a plant form which is found in aquaria, making it very probable that this is an example of contamination of the glass slide before or during processing. (× *400*)

150 Faecal contamination. In addition to amorphous debris, an undigested vegetable fragment is seen in this field. In such a case it is necessary to exclude an anovaginal fistula. (× *250*)

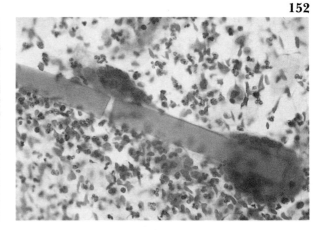

151 Suture material (courtesy of Dr Hampton, St Asaph). Modern absorbable suture material breaks up into short lengths. This smear was taken a few weeks after a repair operation. (× *40*)

152 Suture material with giant multinucleated histiocytes. In this example giant multinucleated histiocytes are acting as foreign-body giant cells. (× *125*)

5 Cervical intraepithelial neoplasia (CIN)

Squamous cell carcinoma of the cervix usually develops in the metaplastic epithelium of the transformation zone; cervical intraepithelial neoplasia (CIN) is seldom seen beyond the last crypt (*Burghardt, 1976*). *Coppleson & Reid (1967)* put forward the theory that immature metaplastic cells are vulnerable to carcinogens (viral or other) which are transmitted at coitus and alter the cell genome to produce a cell line which has neoplastic potential. It may be years after this initiation that other factors trigger progression. It is probable that many cancers go through an intraepithelial phase before becoming invasive, but the cervix is unique in being easily accessible so that women can be screened to detect and treat the disease in the pre-invasive phase. In addition, the complementary discipline of colposcopy can be used to identify the area of abnormal epithelium so that directed biopsies can be taken for histological confirmation.

153

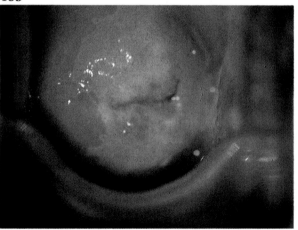

154

153 CIN: colposcopic picture. The reader is referred to standard texts on colposcopy for details of colposcopic examination. This picture shows the cervix painted with 4 per cent acetic acid which causes nuclear-rich epithelium to become densely white. Although this can happen with any immature epithelium, e.g. immature metaplasia, the phenomenon is more striking with CIN when the abnormal area also has a sharp, discrete edge. (*Saline, × 10*)

154 CIN: section of cervix. This is a thick section of cervix mounted and photographed. A vaginal cuff is present, as a hysterectomy was performed in this case for other reasons. Original normal squamous epithelium is reflected from the vagina on to the ectocervix. At the level of the last crypt this becomes normal mature metaplastic epithelium. Approximately half way between this level and the angle of the external os there is a sudden change of epithelial type to a thinner hyperchromatic epithelium which dips into crypts and extends into the lower part of the endocervical canal. (*H&E, × 1.5*)

155 CIN: biopsy of cervix. The sudden change of epithelial type is seen again in this section. Normal squamous epithelium is present in one half of the specimen, and in the other half there is complete failure of differentiation through the full thickness of the epithelium with some flattening on the surface. This corresponds to the original definition of carcinoma-in-situ (*International Committee on Histological Definitions, 1962*) but would now be described as a small-cell undifferentiated CIN III (*Buckley et al., 1982*). (*H&E*, × *80*)

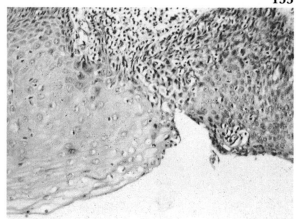

Dyskaryotic cells

Papanicolaou (1954) introduced the term 'dyskaryosis' to describe cells showing nuclear abnormality which ranged from mild changes to changes giving suspicion of invasion. After an interval in which terminology became confused because of variation between laboratories, a Working Party of the British Society for Clinical Cytology (*Spriggs et al., 1978*) recommended a return to the use of 'dyskaryotic' to describe cells shed from lesions ranging from CIN I (mild dysplasia) to lesions giving suspicion of invasion. It was recommended that the term 'malignant cell' should be used only when the smear showed features diagnostic of invasion. In 1986 a further Working Party of the British Society for Clinical Cytology (*Evans et al., 1986*) expanded the 1978 recommendations but supported the use of the term dyskaryosis. For abnormality not amounting to dyskaryosis it was advised that the term 'borderline changes' be used.

Nuclei of dyskaryotic cells are hyperchromatic and at least twice the size of a normal intermediate cell nucleus. The nuclear chromatin shows granular condensation, but without the marked irregularity of clear and condensed areas which is seen in malignant nuclei. The nuclei are usually regular in shape, but bizarre forms can occur. Cytoplasmic differentiation indicates the level of differentiation at the surface of the epithelium from which the cells have exfoliated.

The following table gives an approximation of the anticipated histological lesion depending on the abnormal cells seen in the smear.

Predominant cell type	Dyskaryosis	Anticipated histology
Mature dyskaryotic cells	Mild	CIN I (mild dysplasia)
Small intermediate and parabasal dyskaryotic cells	Moderate	CIN II (moderate dysplasia)
Undifferentiated dyskaryotic cells	Severe	CIN III (severe dysplasia, carcinoma-in-situ, ? invasion)

This is, of course, an oversimplification, as other factors have to be considered, particularly in the diagnosis of CIN III.

The histological diagnosis of CIN III is made in the presence of three epithelial patterns (*Buckley et al., 1982*).

1 Small-cell undifferentiated (see **155**). In smears from such lesions there is a classical picture of small undifferentiated dyskaryotic cells shed singly, in loose clusters, coherent sheets and in tight clusters. In many cases there will be a suggestion of clearing in the nuclei and peaking of the nuclear membrane so that the smear has to be reported as borderline for invasion.

2 Large-cell undifferentiated (see **175**). In these cases dyskaryotic cells in the smear are larger and parabasaloid in type, but single undifferentiated dyskaryotic cells are usually present as well, indicating a more severe lesion than CIN II (see Table). The dyskaryotic parabasal cells may be seen as single cells and as syncytial sheets and clusters with indistinct cell borders. It may be necessary to report the smear as suggestive of CIN II/III.

3 Differentiated (large-cell keratinizing) (see **178**). Tissue sections show complete disorganization of growth pattern and cells have a relatively low nucleocytoplasmic ratio; keratinization and surface keratosis are common. In the cervical smear many cells will present as well-differentiated dyskaryotic cells, but the nuclear abnormality will be more severe than that seen in CIN I. In addition, single undifferentiated dyskaryotic cells are usually found. Abnormal cytoplasmic differentiation taking the form of 'fibre cells', 'tadpole cells' and other bizarre forms is often present and when there is nuclear degeneration the cellular picture may come into the category 'borderline for invasion'.

Although the nuclei of dyskaryotic cells are usually hyperchromatic, in some cases, particularly in cases of CIN III (large- and small-cell), the nuclear chromatin may be pale staining but it is usually possible to recognize the coarse granularity of the chromatin pattern. Problems also arise when dyskaryotic endocervical columnar cells are present, as these can overlie tissue which will be reported as CIN III on histological section.

The photographs in this section illustrate the wide range of cell patterns seen in the presence of CIN.

156

157

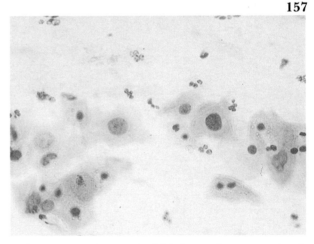

156 Dyskaryotic and reactive cells. The intermediate cells in this field have slightly enlarged but otherwise normal nuclei. Contrast these with the denser nuclei seen in two parabasal cells and one small intermediate cell. These three cells are eosinophilic and mildly dyskaryotic. (× *160*)

157 Mild dyskaryosis. This is another example of both reactive, amphophilic staining cells and mild dyskaryosis. (× *160*)

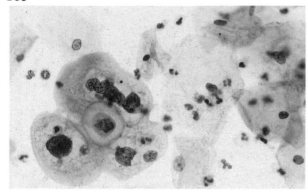

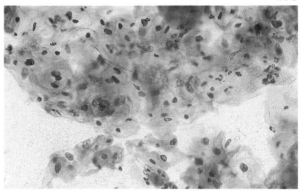

169 Mild dyskaryosis and HPV infection. At
higher magnification granular condensation is seen
in the multinucleated cell and in the early koilocyte.
(× *160*)

**170 Mild dyskaryosis with probable HPV
infection.** Dyskaryotic cells and cells showing
degenerative changes are seen among keratinized
cells. No koilocytes are present, but it is probable
that this is a case of HPV infection. Although the
appearances suggest only CIN I, cells reflecting
surface keratosis make colposcopy and biopsy
advisable. (× *100*)

Examples of cases

Case 1: CIN III

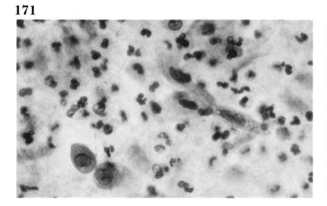

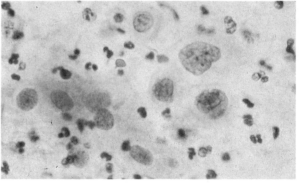

171 Keratinized dyskaryotic cells. This is an
infected smear with some cell degeneration.
Keratinized dyskaryotic cells are seen, two of which
are 'fibre cells'. (× *125*)

172 Parabasal dyskaryotic cells. Another field
from the same smear shows a small cluster of
parabasal dyskaryotic cells. The staining is light, but
the granular condensation of chromatin can be
compared with the normal intermediate cells in the
same smear. This may result from hydropic
degeneration. (× *250*)

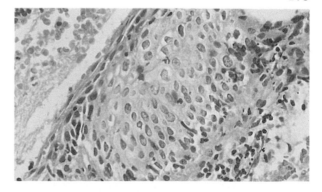

173 CIN III: endocervical curetting. This
fragment of epithelium was reported as CIN III.
Note the elongated keratinized cells of the surface
layers and the failure of differentiation in the
remaining thickness of the epithelium. The nuclear
staining is pale, as in the cervical smear.
(*H&E*, × *125*)

Case 2: CIN III

174

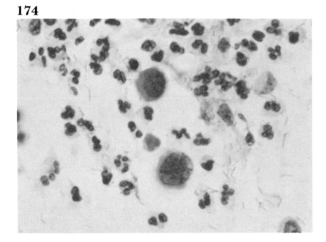

175

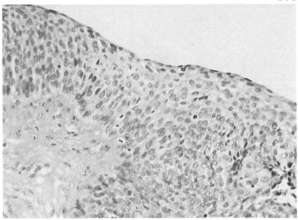

174 Infected smear with dyskaryotic cells. This is an infected smear consisting mainly of polymorphs, but occasional undifferentiated and parabasal dyskaryotic cells were scattered throughout the smear. The staining is blurred and amphophilic because of infection, but the granular condensation of the chromatin is still apparent. (× 250)

175 CIN III: punch biopsy of cervix. This is again a large-cell undifferentiated CIN III. The nuclear staining is pale, but note the presence of mitoses. (H&E, × 62)

Case 3: CIN III

176

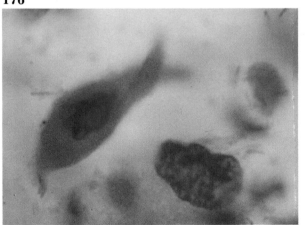

176 Degenerative dyskaryotic cells. This field contains a dyskaryotic nucleus with frayed cytoplasm and a bizarre keratinized cell. Both cells show nuclear degeneration. (× 620)

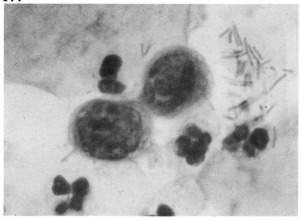

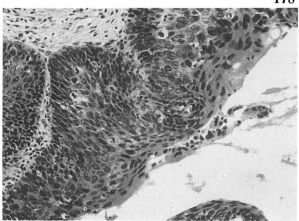

177 Severe dyskaryosis. Another field in the same smear shows poorly differentiated dyskaryotic cells with severely dyskaryotic nuclei which show a suggestion of clear areas in the nucleus. This could be caused by degeneration, but must raise suspicion of invasion. (× 620)

178 CIN III: cone biopsy. Sections from the cone biopsy show CIN III of the keratinizing type. It will be noted that the surface layers consist mainly of bizarre keratinized cells with a disorganized growth pattern. Capillaries in stromal papillae bring the undifferentiated cells of the deeper layers close to the surface, thus accounting for their presence in the smear. There was no evidence of invasive carcinoma. (*H&E*, × 62)

Case 4: CIN III with HPV infection

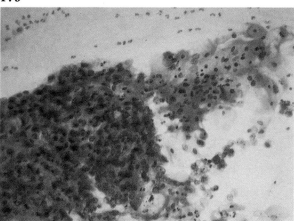

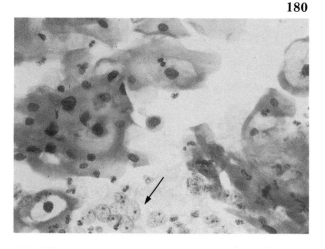

179 Dyskaryotic and keratinized cells. In addition to a sheet of keratinized and dyskeratotic cells, this field contains a koilocyte and a dense cluster of poorly differentiated dyskaryotic cells. Pale nuclei with prominent nucleoli are seen, which suggest dyskaryotic endocervical columnar cells. (× 100)

180 Pale-staining dyskaryotic cells and glandular nuclei. These features are seen again at higher magnification in another field from the same smear. Note the single parabasal dyskaryotic cell in the centre of the field. Note the cluster of glandular nuclei which are pale-staining and have prominent nucleoli (arrow). (× 160)

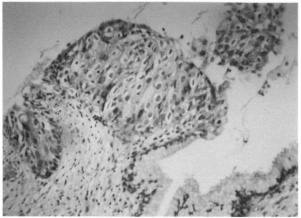

181

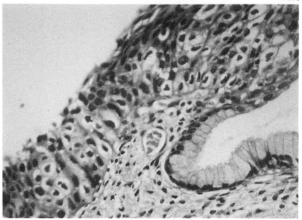

182

181 CIN III with HPV infection: cone biopsy. Abnormal epithelium is seen at a crypt neck. Note the dyskaryotic columnar cells on the surface at one edge. There is loss of polarity in the tissue, and this was diagnosed as CIN III, but the presence of koilocytes throughout the epithelium makes it difficult to assess the degree of CIN (*Fletcher, 1983*). (*H&E, × 100*)

182 CIN III with HPV infection: cone biopsy. This is seen more clearly at higher magnification in another field. (*H&E, × 160*)

Case 5: CIN III (fusiform)

183

184

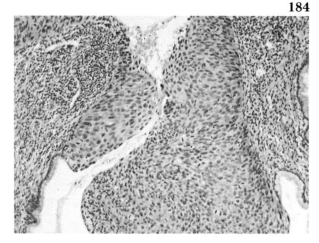

183 Fusiform cells. This smear consists of dense clusters of cells with hyperchromatic nuclei. The photograph is taken at high magnification at the edge of the cluster to show that the cells are fusiform, so the possibility of a fibroma or fibrosarcoma arises. (× 620)

184 CIN III (fusiform): cone biopsy. Tissue sections show the typical picture of a small-cell undifferentiated CIN III with gland involvement. (*H&E, × 37.5*)

185 CIN III (fusiform): cone biopsy. However, when the tissue is examined at the same magnification as the cervical smear it can be seen that the cells are fusiform. (*H&E, × 620*)

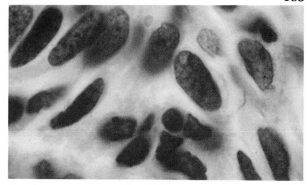

Case 6: CIN III with HPV infection

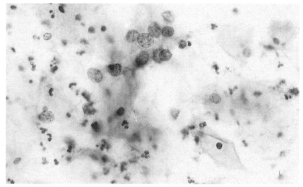

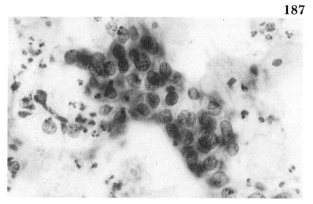

186 Keratinized and dyskaryotic cells. An infected smear with keratinized cells and some degenerate dyskaryotic squames. Endocervical columnar cells are also present. (× 160)

187 Dyskaryotic cells. Another field in the same smear shows a sheet of poorly differentiated dyskaryotic cells. There is a community border on one surface, which suggests that these are columnar cells covering the deeper cells which could reflect reserve-cell hyperplasia or CIN III. (× 160)

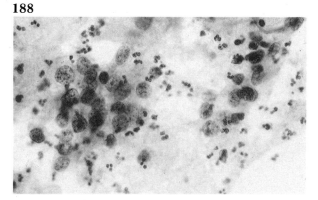

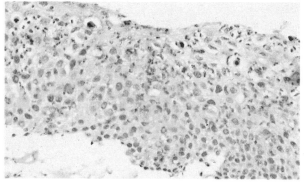

188 Pale, undifferentiated dyskaryotic cells. Another field in the same smear shows a loose cluster of undifferentiated dyskaryotic cells. Staining is pale, but granular condensation of chromatin is seen. (× 160)

189 CIN III with HPV infection: colposcopic biopsy. The colposcopic biopsy from this case was reported as showing CIN III with occasional koilocytes present. There is disorganization of growth pattern with pleomorphism and individual cell keratinization, as well as occasional koilocytes. Polymorphs have infiltrated the epithelium and it is difficult to assess the extent to which inflammatory reaction is influencing the apparent degree of CIN. Note the equivalent pallor of freshly biopsied intact epithelium compared with the cells in the cervical smear. (*H&E, × 80*)

Case 7: CIN II

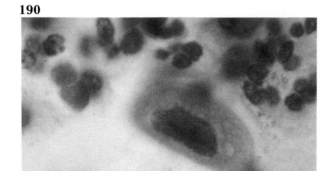

190 Well-differentiated dyskaryotic cells.
Well-differentiated dyskaryotic squamous cells were present in this smear. The example shown is a tadpole form. (× 620)

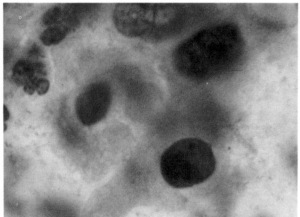

191 Parabasal dyskaryotic cells. Elsewhere in the smear parabasal dyskaryotic cells were seen. (× 620)

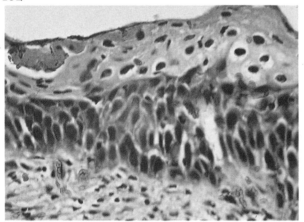

192 CIN II: cone biopsy. In this section there is failure of differentiation which extends into the middle third of the epithelium and continues with abnormal differentiation to the surface. Note the horizontal Schiller line. In cases such as this the superficial layers are detached easily if care is not taken in handling the specimen. This can result in a diagnosis of CIN III and an apparent failure of cytological—histological correlation. (*H&E*, × 125)

Case 8: CIN I/II

193 Dyskaryotic cells. In this case low magnification shows profuse exfoliation of mild-to-moderate dyskaryotic cells. Some cells are keratinized but koilocytes were not found. (× *40*)

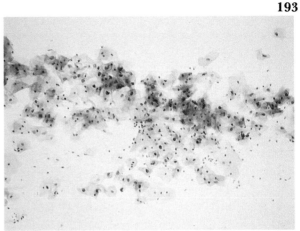
193

194 CIN I/II with koilocytosis: cone biopsy. This section from the cone biopsy shows the range of CIN in the epithelium. Note the presence of koilocytes in the superficial layers. (× *40*)

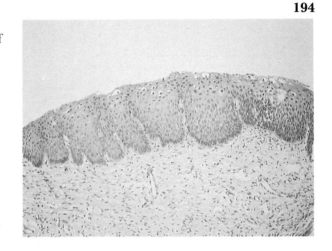
194

195 CIN I/II with koilocytosis. At higher magnification the sharp transition between an area of CIN I and an area of CIN II is seen. Note the capillary between these areas (arrow), which takes deeper layers near the surface so that occasional severely dyskaryotic cells might have been expected in the cervical smear. (*H&E*, × *80*)

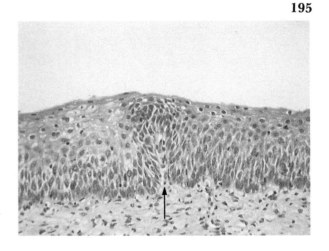
195

6   Invasive cancer of cervix

196

197

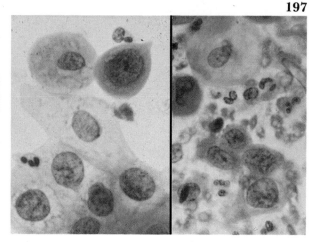

196 Squamous cell carcinoma of cervix: colposcopy. This woman had a procidentia and it was not clear to the naked eye whether this was traumatic ulceration or an invasive cancer. Magnification shows the undulant surface and vascular irregularity. Contact bleeding is also present. (*Saline, green filter, × 10*)

197 Normal, dyskaryotic and malignant cells. This composite photograph shows cells with normal and dyskaryotic nuclei on the left and cells with normal and malignant nuclei on the right. Compare the nuclear chromatin pattern of the parabasal dyskaryotic cell with that seen in the cluster of poorly differentiated malignant cells. The most striking difference is the irregular clear areas seen between clumped chromatin in the malignant cells. In addition, the malignant nuclei show irregularly shaped nucleoli, and one nucleus illustrates marked peaking of the nuclear membrane. (× 250)

Criteria of malignancy

When non-gynaecological cytology preparations are reported it is necessary to distinguish between normal, reactive and malignant cells. In gynaecological reporting there is the additional complication of recognizing dyskaryotic cells, which is the need to distinguish between an intraepithelial and an invasive lesion. The reliability of the report depends on recognition of good criteria of malignancy, and in many cases the diagnosis can be as accurate as with a tissue section. However, it has to be accepted that there are cases in which the criteria of malignancy are insufficient, although suspicious features may be present. In these cases the abnormal cells should be described as dyskaryotic with a supplementary statement outlining the reasons for suspicion of an invasive lesion. This is important, as it will modify clinical management of the case, particularly with regard to size and depth of biopsy.

The most useful criteria of malignancy are nuclear and nucleolar changes, relationship of nucleus to cytoplasm and relationship of one cell to another. Other features such as multinucleation, hyperchromatism, cell shape and pleomorphism have to be applied with caution. It has already been shown that these changes may reflect only reaction and cell regeneration.

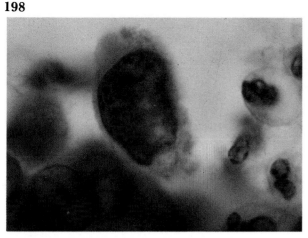

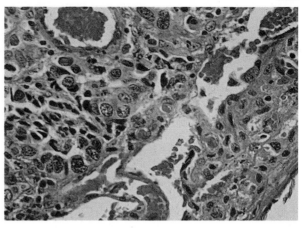

198 Malignant nucleus: chromatin clumping.
In dyskaryotic cells there is coarse clumping of the
chromatin, but the chromatin is uniformly
distributed throughout the nucleus. In comparison,
the nucleus in this field shows large, irregular
clumps of chromatin with irregular, clear areas
between them. It is important to be certain that the
chromatin clumps have a sharp edge, as coagulative
necrosis gives a similar appearance except for
diffuse, blurred edges. This also applies to irregular
condensation of chromatin at the nuclear
membrane, a feature also seen in this nucleus.
Irregularities of the nuclear membrane are also
important and a sharp indentation with peaking at
one pole is seen in this nucleus. (× 620)

199 Squamous cell carcinoma: wedge biopsy.
Examination of tissue from the invasive cancer in
this case shows that the same irregularities of
nuclear chromatin are present. (H&E, × 125)

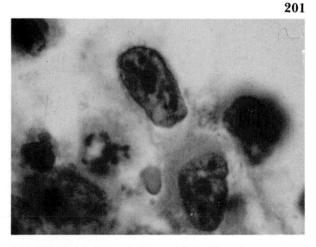

200 Squamous cell carcinoma: wedge biopsy.
This is more apparent when the tissue is
photographed at the same magnification as the
cytological preparation. Note the irregularly shaped
nucleoli. (H&E, × 620)

201 Malignant nucleus: nuclear membrane. In
this example there is irregularity of condensed
chromatin at the nuclear membrane. At one pole it is
very thick and at the other it is absent. Note also the
irregularly shaped nucleoli and the spikiness of
nuclear chromatin and nucleolar membrane. (× 620)

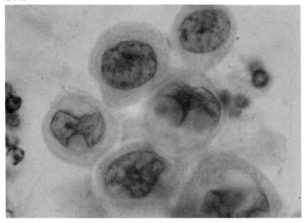

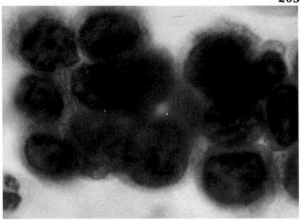

202 Malignant nuclei: irregularities of nuclear shape. In addition to irregularities of the chromatin, two of these nuclei show sharp, clear-cut indentations into the nucleoplasm; there is also some vacuolization. The other nuclei show finer irregularities of the nuclear membrane. (× *620*)

203 Coagulative necrosis: degenerative dyskaryotic nuclei. This field illustrates the way in which confusion arises when coagulative necrosis is present. Comparison with **198-202** shows that in this picture the edges of chromatin clumps are diffuse and the indentation into the nucleoplasm is smooth and rounded. (× *620*)

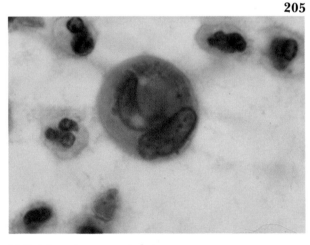

204 Keratinized malignant cell. This cell shows two features of interest. The nuclei are placed to the side of the cell and the cytoplasmic border exactly parallels the nuclear border of the larger nucleus; this is another feature seen in malignant cells (*Frost, 1969*). The cell is binucleate, but the nuclei differ markedly in size and one is vesicular, whereas the other is pyknotic. Irregular keratinization is also present, but this finding is non-specific. (× *620*)

205 Binucleate malignant cell. This is another example of a binucleate cell with nuclei of different shapes and relationship to each other. Contrast this with previous examples of multinucleate cells seen in reaction and dyskaryosis (**72, 73, 85, 104, 158, 169**). (× *620*)

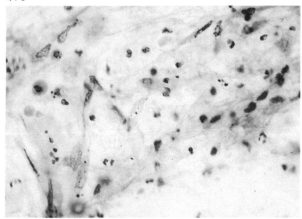

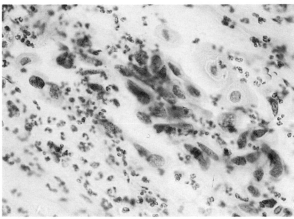

206 Malignant fibre cells. At a lower magnification this field shows a number of degenerate keratinized fibre cells. Comparison with **74** shows that these differ in the marked irregularities of nuclear chromatin which are seen. It is this feature which leads to a diagnosis of malignancy, and not the shape of the cell. (× *160*)

207 Fusiform malignant cells. Comparison with **80** and **183** shows that in this case the nuclear chromatin is irregularly clumped with clear areas. Pleomorphism and irregularities of nuclear outline are also present. Degenerative changes are less marked than in **206**. (× *160*)

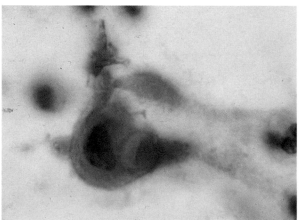

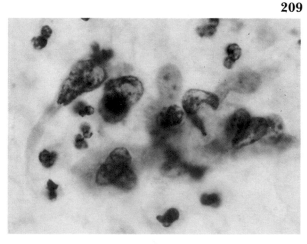

208 Malignant tadpole. This is an example of abnormal differentiation in a malignant cell. Note that the tail seems to be in the process of being pinched off from the rest of the cell. This suggests another example of abnormal division with lysis of the second nucleus. In other fields and on biopsy this was confirmed as a squamous cell carcinoma, but this small single cell shows too much nuclear degeneration for a reliable diagnosis to be made. Comparison with the polymorph nucleus in the field shows that this is a small cell. (× *620*)

209 Malignant tadpole. By contrast, in this field, which also shows examples of abnormal differentiation, nuclei show good criteria of malignancy. (× *400*)

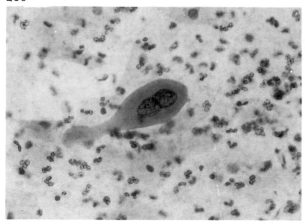

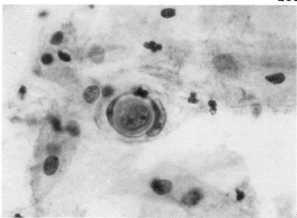

210 Dyskaryotic tadpole. Compare the nuclear chromatin pattern of this binucleate dyskaryotic cell with the nuclear chromatin pattern in **209**. It will be seen in this example that the nuclear chromatin pattern resembles that seen in the dyskaryotic cells illustrated in Chapter 5. Note also that the two nuclei are identical, unlike the paired nuclei seen in **204** and **205**. (× *160*)

211 Malignant pearl. Compare this example of pearl formation with **82** and **83**. It will be seen that in comparison with a benign and a dyskaryotic pearl, this can be recognized as a malignant pearl because of the more irregular nuclear chromatin pattern. (× *320*)

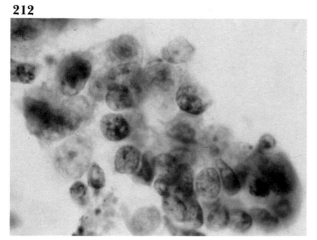

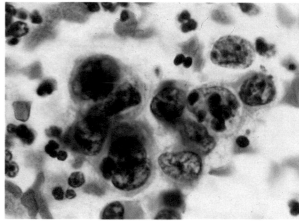

212 Malignant columnar cells. The examples of malignant cells already illustrated have been from squamous cell carcinomas, and in most cases cytoplasmic differentiation has been in keeping. In this field acinar forms are seen and the cytoplasm is floccular. It will be noted that the nuclear chromatin is granular rather than clumped, and the irregular areas of clearing are smaller but still present. Nucleoli are more prominent. (× *400*)

213 Undifferentiated malignant cells. This group of malignant cells has a high nucleocytoplasmic ratio and it is difficult to be sure of the type of cytoplasmic differentiation. The sharp cytoplasmic border and concentric ribbing in one cell make it most likely that this is a small-cell squamous carcinoma, but it would probably be wise to describe it as 'differentiation uncertain'. (× *400*)

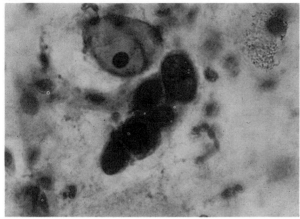

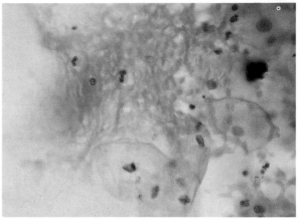

214 Undifferentiated malignant cells. This is another example of malignant cells of uncertain differentiation. In this case histological sections were also reported as an undifferentiated carcinoma. (× *250*)

215 Tumour diathesis. This pictures shows the sero-sanguinous background which has been described as a 'tumour diathesis'. Alone this is a non-specific finding, as it can be seen in the presence of tissue damage from any cause. (× *40*)

Examples of cases with squamous cell carcinoma

Case 1

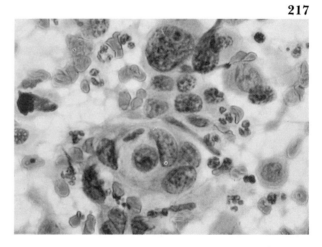

216 Malignant squamous cells. Malignant squamous cells are seen and one binucleate cell has a smaller keratinized cell superimposed on it. Note the tendency to concentric ribbing of the cytoplasm; this supports a diagnosis of squamous cell carcinoma. (× *250*)

217 Malignant pearl. In another field there is a good example of a malignant pearl. (× *250*)

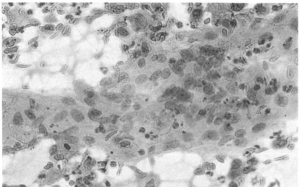

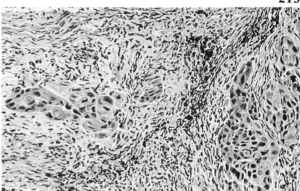

218 Syncytium. At a lower magnification in another field, a syncytium of paler, poorly staining cells is seen. These are not obviously malignant, and could reflect a histiocytic reaction. (× 125)

219 Squamous cell carcinoma: cervical biopsy. The biopsy from this case shows tongues of malignant tissue in a reactive stroma. (*H&E*, × 62)

Case 2

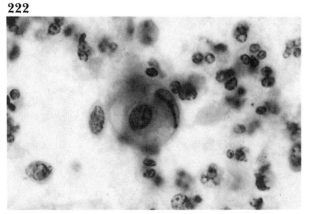

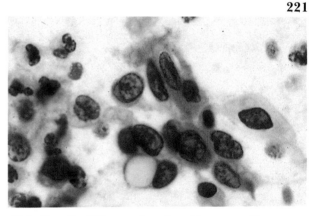

220 Degenerate malignant cell. In this field a single well-differentiated malignant cell is seen, showing nuclear degeneration and peripheral fraying of the cytoplasm. (× 400)

221 Poorly differentiated malignant cells. Elsewhere, the malignant cells are better preserved but less well-differentiated. (× 400)

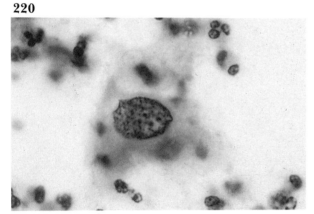

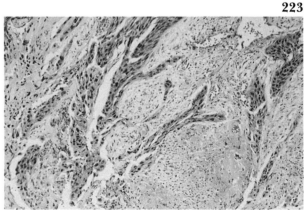

222 Nuclear moulding. This group is shown in order to demonstrate moulding of the nucleus of one cell to the cytoplasm of another and early pearl formation. (× 400)

223 Squamous cell carcinoma: cervical biopsy. The biopsy shows tumour infiltrating from a crypt. (*H&E*, × 40)

224 Squamous cell carcinoma: cervical biopsy.
At higher magnification the pleomorphism seen in the cervical smear is illustrated. (*H&E, × 160*)

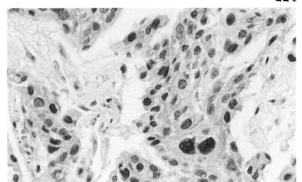

Case 3: Squamous cell carcinoma with herpesvirus infection

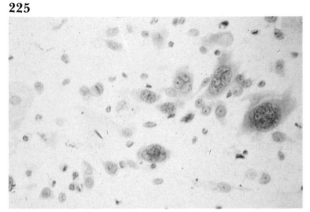

225 Herpes changes. This case is included as an interesting example of herpes virus infection superimposed on a squamous cell carcinoma. In this low-power field single and multinucleate cells are seen with intranuclear inclusions characteristic of herpes, but it is difficult to be sure whether these cells are also malignant. (*× 125*)

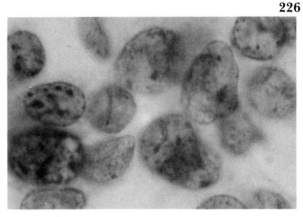

226 Malignant cells. At a high magnification single malignant cells are seen which have ground-glass nuclei in some cells and clumping of nuclear chromatin in others. Nucleoli are also present. (*× 620*)

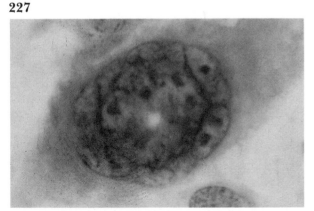

227 Multinucleated cell. Compare this cell with that seen in **126**. Both show close-packed nuclei with intranuclear inclusions. In both cases the appearances suggest no more than herpesvirus infection. (*× 620*)

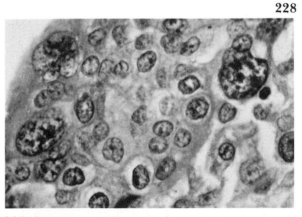

228 Squamous cell carcinoma with herpesvirus infection: cervical biopsy. Tumour tissue from this case shows a similar range of malignant cells, but typical herpes changes are not seen. The presence of herpesvirus infection was confirmed in this case. (*H&E, × 250*)

Case 4: Recurrence after radiation

229

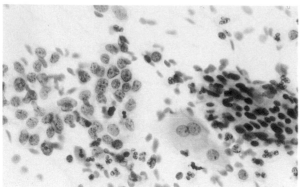

230

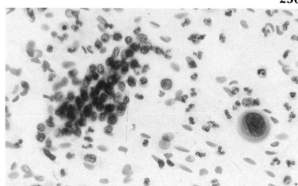

229 Malignant and dyskaryotic cells. This patient had had a squamous cell carcinoma of the cervix treated with radium 6 months previously. This field shows small, undifferentiated malignant cells and a sheet of darker dyskaryotic cells. (× *160*)

230 Malignant cells and radiation effect. Another field shows more small, undifferentiated malignant cells with a large parabasal cell which shows changes, probably due to radiation. (× *160*)

231

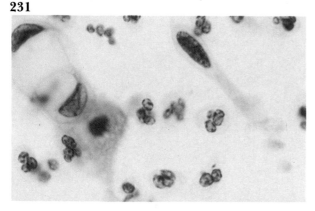

231 Malignant tadpole. The original tumour was a well-differentiated keratinizing squamous carcinoma, but the tadpole shown in this field was one of the few differentiated cells in the smear. It is not uncommon for tumours recurring after treatment to become poorly differentiated. As this patient had an obvious clinical recurrence, no tissue is available. (× *400*)

Case 5: Recurrence after radiation

232

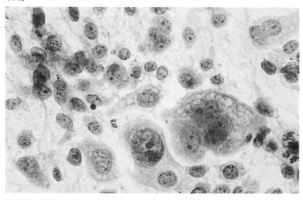

233

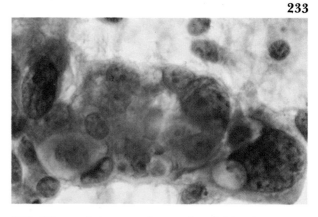

232 Pleomorphic cells. This is another example of recurrence after radiotherapy, showing a pleomorphic picture of large undifferentiated cells with prominent nucleoli and cytoplasmic vacuolation. One multinucleated cell is present. (× *125*)

233 Tissue fragment. In another field a tissue fragment is seen which consists of similar cells. (× *125*)

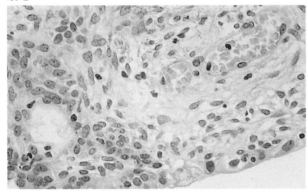

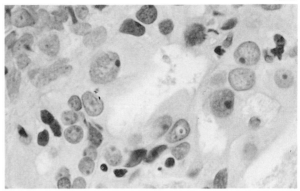

234 Recurrence after radiation: biopsy of polyp. In this case recurrence was in an endocervical polyp and this section shows rather palely staining tumour cells infiltrating the polyp. Note the remaining columnar cells on the surface. (*H&E, × 125*)

235 Recurrence after radiation: biopsy of polyp. At higher magnification the tumour cells in the tissue can be compared with those seen in the cervical smear. (*H&E, × 250*)

Examples of cases with adenocarcinoma of the cervix

There are three main types of adenocarcinoma of the cervix: the tumour pattern can be endocervical, endometrial or clear cell (mesonephroid) in type.

Case 1

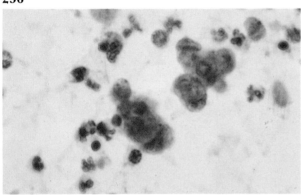

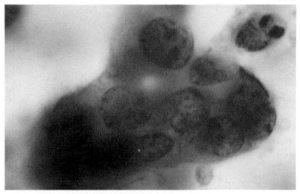

236 Poorly differentiated malignant cells. Clusters of poorly differentiated malignant cells are seen in this field. (*× 250*)

237 Poorly differentiated malignant cells. At very high magnification cytoplasm is seen and seems to be of the endocervical type. (*× 620*)

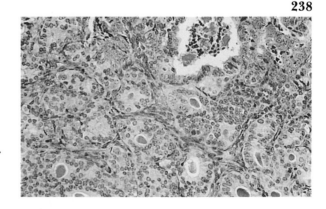

238 Adenocarcinoma of cervix: cervical biopsy. It is interesting that sections from the cervix show a poorly differentiated tumour of the clear cell (mesonephroid) type. (*H&E, × 62*)

Case 2: Adenocarcinoma of cervix—clear cell

239

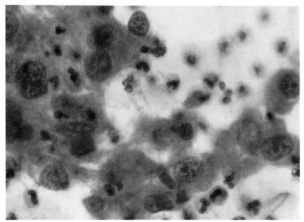

239 Malignant cells. This smear was taken from a woman aged 24 years. It shows a pleomorphic picture of malignant cells of the clear cell type. (× *250*)

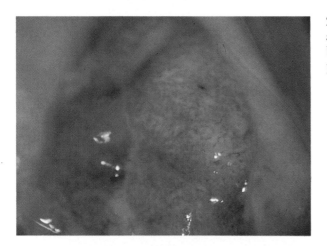

240 Adenocarcinoma of cervix: colposcopy. At a very high magnification the tumour can be seen to bulge from the cervix. Note the irregular vascular pattern. (*Saline,* × *40*)

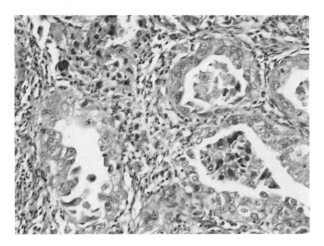

241 Adenocarcinoma of cervix: Wertheim hysterectomy. Sections showed an adenocarcinoma of mesonephroid type. One gland shows a hobnail pattern. (*H&E,* × *62*)

Case 3: Adenocarcinoma—endocervical type

242

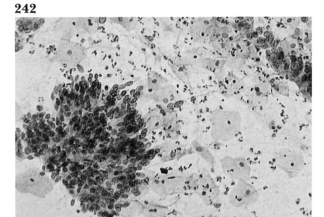

243

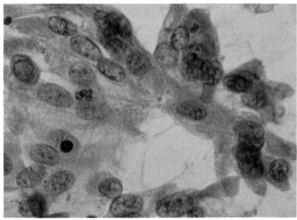

242 Malignant endocervical columnar cells. It will be seen that at low magnification the tissue fragment resembles the appearance that would be seen if these were reactive endocervical columnar cells. (× 62)

243 Malignant endocervical columnar cells. At higher magnification it can be seen that the nuclei show good criteria of malignancy (compare with **78**). This diagnosis was confirmed on tissue section. (× 250)

Miscellaneous cases

Carcinoma of cervix can be both adeno and squamous in type. In these cases either or both elements may be in-situ or invasive.

Case 1: Combined adenosquamous carcinoma (courtesy Dr J.K. Frost, Johns Hopkins Hospital)

244

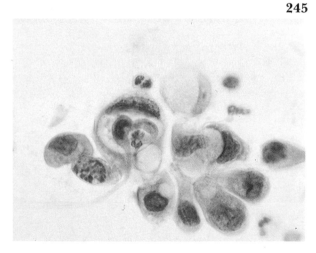

245

244 Malignant squamous cells. This field contains a cluster of malignant squamous cells with one mildly dyskaryotic cell. (× 280)

245 Acinar fragments. Malignant acinar fragments are seen in another field from the same smear. A tadpole overlies one of the acini. Both invasive squamous cell carcinoma and adenocarcinoma were present on biopsy. (× 280)

Case 2: Adenocarcinoma with CIN III
(courtesy Dr A. Kertesz, Merthyr Tydfil)

246

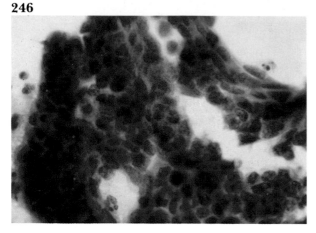

247

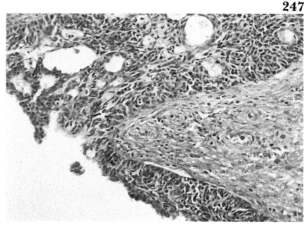

246 Severe dyskaryosis. This field shows sheets and clusters of severely dyskaryotic cells, but there is also a suggestion of papillary structures. (× *250*)

247 Adenocarcinoma with CIN III: cone biopsy. Cone biopsy showed areas of CIN III. In this field residual benign columnar cells are present on the surface. (*H&E, × 62*)

248

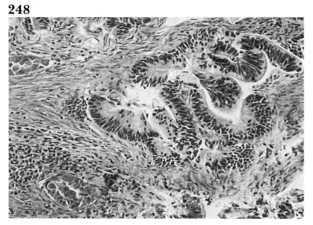

248 Adenocarcinoma with CIN III: cone biopsy. Other sections showed a well-differentiated adenocarcinoma of the endocervical type. (*H&E, × 62*)

Case 3: Adenocarcinoma-in-situ

This case is included as the cytological appearances show many of the features considered to be diagnostic of adenocarcinoma-in-situ (*Ayer et al., 1987*).

249

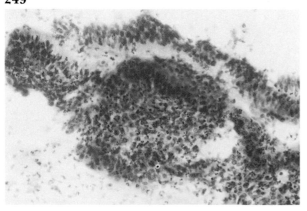

249 Endocervical gland. This field shows a gland in profile with pseudostratification. (× *80*)

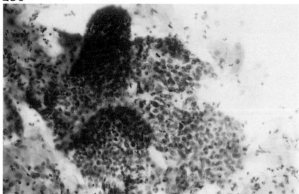

250 Endocervical gland. In this picture surface columnar epithelium has exfoliated with finger-like glands pulled from the stroma. (× *80*)

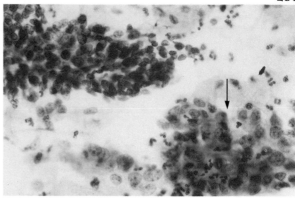

251 Dyskaryotic columnar cells. Elsewhere in the smear a sheet of undifferentiated dyskaryotic cells can be contrasted with a sheet of dyskaryotic columnar cells. Note the prominent nucleoli and paler nuclei in the glandular cells (arrow). (× *160*)

252

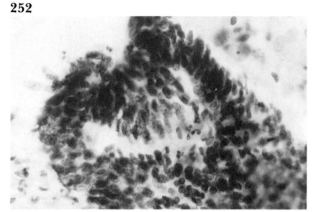

252 Feathered edge. This field shows a gland opening with pseudostratification. Note the feathered edge seen on one surface. *Ayer et al.* consider this to be an important diagnostic feature. (× *160*)

253

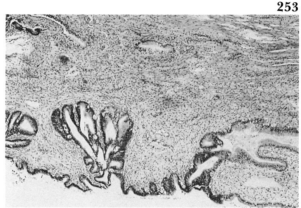

253 Adenocarcinoma-in-situ: cone biopsy. This section shows a well-differentiated adenocarcinoma-in-situ. (*H&E, × 16*)

254

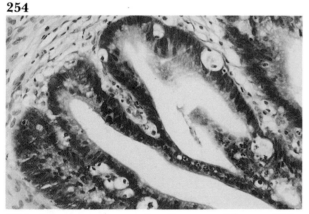

254 Adenocarcinoma-in-situ: cone biopsy. This is seen more clearly at higher magnification. (*H&E, × 160*)

255

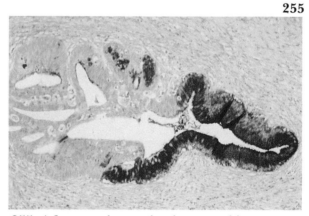

255 Adenocarcinoma-in-situ: cone biopsy. Areas of squamous metaplasia are present which show foci of residual mucus staining red with PAS–diastase. The adenocarcinoma-in-situ component is also strongly positive for mucus. (*PAS–diastase, × 160*)

Case 4: Malignant melanoma of cervix (courtesy Edgware General Hospital)

256

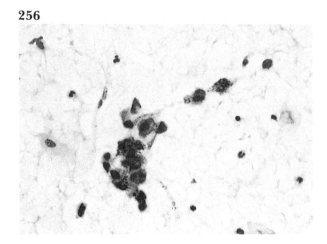

256 Malignant melanoma. Pigment is seen in the cell cluster in this field. (× *160*)

257

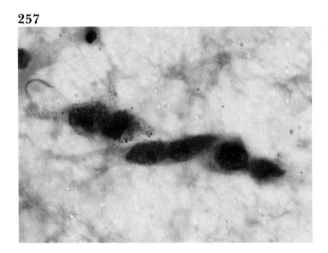

257 Malignant melanoma. At higher magnification irregularities of the nuclear chromatin pattern can be seen through the heavy pigment deposition. In particular, note the faintly discernible macronucleoli; these can be an important diagnostic feature in cases of amelanotic melanoma (see **336-339**). (× *400*)

258

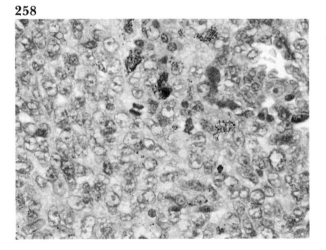

258 Malignant melanoma: cervical biopsy. Compare the constituent cells in the tissue with those seen in the cervical smear. Note the macronucleoli. (*H&E*, × *160*)

Case 5: Basi-squamous carcinoma of cervix

259 Poorly differentiated malignant cells. A cluster of poorly differentiated malignant cells is seen in this field. They show no features to suggest squamous cell or columnar cell differentiation. (× 250)

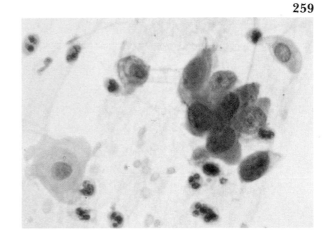

260 Basi-squamous carcinoma of cervix: cervical biopsy. At low magnification nests of tumour cells are seen in the stroma. This was diagnosed as a basi-squamous cell carcinoma. (H&E, × 18)

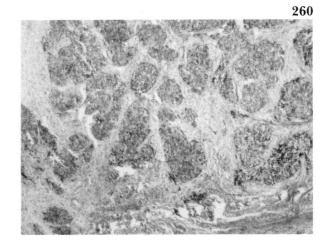

261 Basi-squamous carcinoma of cervix: cervical biopsy. At higher magnification the tumour cells can be compared with those seen in the cervical smear. Note the darker-staining endothelial cells. (H&E, × 250)

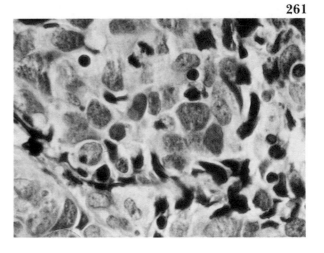

Metastatic carcinoma

Metastatic carcinoma is diagnosed occasionally when the symptoms presented by the patient relate to the secondary rather than the primary tumour. The true incidence is probably higher than is apparent, because it is not usual to include the cervix in a search for metastatic tumour from a known primary.

Case 1: Metastatic tumour from pancreas (courtesy Dr J.K. Frost, Johns Hopkins Hospital)

262

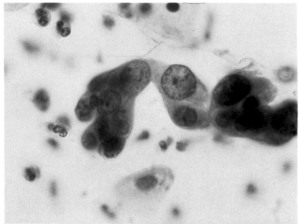

262 Malignant cells. These unusual malignant cells were found in a smear, taken as a screening smear, from a woman who was later found to have a carcinoma of the pancreas. (× *280*)

Case 2: Metastatic tumour from breast

263

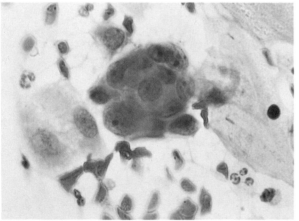

263 Morula cluster. This woman was known to have had a carcinoma of the breast some years previously. The morula cluster found in a routine cervical smear resembled clusters found in some cases of fine-needle aspiration of breast cancer. The presence of a secondary deposit was confirmed by biopsy. (× *250*)

7 Endometrial cytology

It is normal to see endometrial cells in cervico-vaginal smears in relation to menstruation, and examples are shown in Chapter 1. In the presence of adenocarcinoma of endometrium profuse exfoliation of diagnostic malignant cells occurs in, perhaps, 20 per cent of cases; in another 20 per cent there may be no exfoliation and no features to cause suspicion of endometrial pathology. In the remaining majority, although diagnostic cells are not seen, there may be features which indicate the need for further investigation. The final diagnosis must be histological.

Exfoliation of malignant endometrial cells can be very sparse, so meticulous screening of such smears is essential. For this reason many laboratories make it their practice to double-screen all smears when there is a clinical history of irregular or postmenopausal bleeding.

The following are features which should cause suspicion of endometrial pathology.

1 The presence of normal endometrial cells after the menopause and, to a lesser extent, normal endometrial cells not related to menstruation in a woman who is still cyclical.

2 A cell pattern showing an oestrogen effect in a postmenopausal woman. This can also be caused by an ovarian or breast tumours.

3 The presence of dyskaryotic endometrial cells. In many cases of well-differentiated adenocarcinoma, cells found in the smear may show no more abnormality than would be expected with reaction or hyperplasia.

Diagnostic problems

Case 1: Sparse exfoliation

264

265

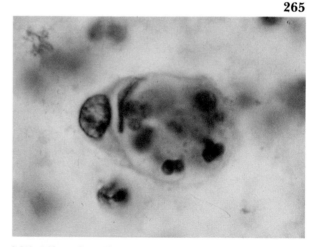

264 Missed malignant fragment. This field illustrates an infected smear with many polymorphs present. At screening magnification the malignant fragment at the centre of the field is difficult to see and was missed on the original screening. Meticulous rescreening showed no other malignant cells. (× 125)

265 Missed malignant fragment. At high magnification this can be identified as a malignant fragment. (× 620)

Case 2: Oestrogen effect and endometrial debris

266

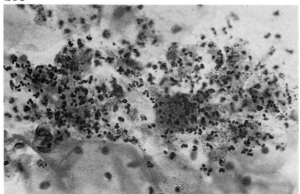

266 Endometrial debris. This smear was taken from a woman with postmenopausal bleeding. An oestrogen effect is shown by the cell pattern but, in addition, there is a streak of amorphous debris with degenerate polymorphs and other cells. This feature has been reported in cases of endometrial cancer, but should be treated with caution as the presence of debris and necrosis reflects tissue breakdown which can also occur with severe endometritis or ulceration. In this example curettage confirmed the presence of adenocarcinoma of endometrium. (× 125)

Case 3: A case of endometritis for comparison

267

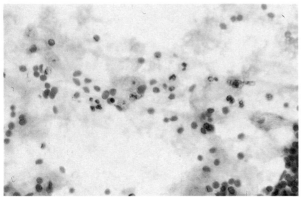

267 Endometrial debris. This field shows a similar but less marked picture. There is a serous background with red cells and degenerate reactive cells with prominent nucleoli. (× 160)

268

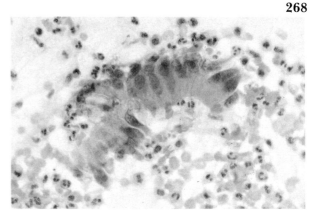

268 Degenerative columnar cells: endometrial aspirate. This field from the endometrial aspirate contains a strip of degenerate hyperplastic columnar cells. (× 160)

269

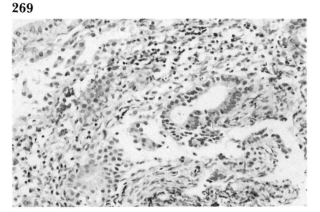

269 Endometritis: curettings. Curettage showed endometritis only. Note the scanty glands, some of which are disrupted, and the stroma infiltrated by inflammatory cells. (H&E, × 80)

270

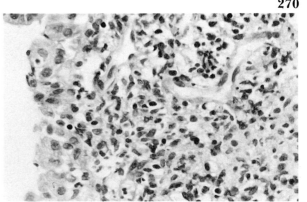

270 Endometritis: curettings. At higher magnification the stroma is seen to be vascular with a lymphocytic reaction which infiltrates the surface columnar epithelium. (H&E, × 160)

Similarities between malignant and reactive endometrial cells

These paired examples demonstrate the difficulties in establishing the degree of anticipated abnormality from the appearance of atypical endometrial cells as seen in cervicovaginal smears. It has to be remembered that the cells have travelled from the cavity of the uterus, in menstrual or serous fluid, and some degree of degeneration would be expected. In each of these pairs the example of adenocarcinoma is presented first. In all of these cases curettage was essential to establish the diagnosis.

Pair 1

271

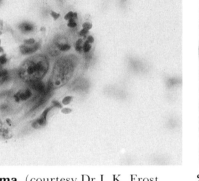

272

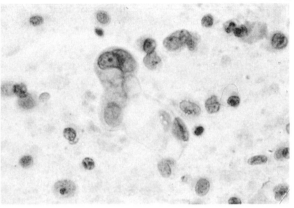

271 Adenocarcinoma. (courtesy Dr J. K. Frost, Johns Hopkins Hospital). (× *280*)

272 Reaction to IUCD. This tissue fragment shows cytoplasmic vacuolization, and prominent nucleoli are present. However, there is no irregularity of nuclear chromatin. (× *250*)

Pair 2

273

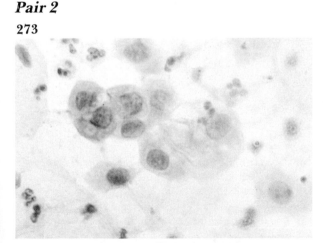

274

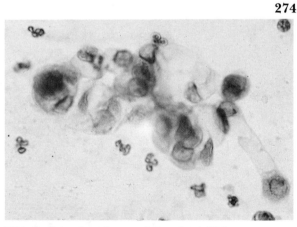

273 Adenocarcinoma. The nuclei in this group of cells are no more than dyskaryotic, but adenocarcinoma of the endometrium was found on curettage. (× *250*)

274 Infected endometrial polyp. This photograph demonstrates the caution needed when nuclear degeneration is present. Coagulative necrosis makes the nuclear chromatin pattern look more abnormal than the cells in **273**. This woman had an endometrial polyp which presented through the cervical os. No evidence of malignancy was found. (× *250*)

Pair 3

275

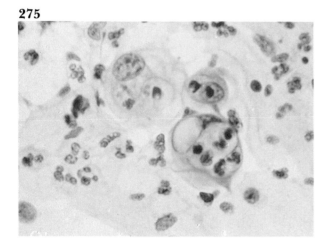

276

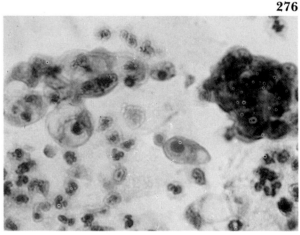

275 Adenocarcinoma. Nuclei in this fragment show an irregular chromatin pattern as well as cytoplasmic vacuolation, with polymorphs in the vacuoles. (× 250)

276 Cystic hyperplasia of endometrium. Curettage showed only cystic hyperplasia in this woman, but it will be seen that nuclear abnormality is as marked as that seen in **275**. (× 250)

Pair 4

277

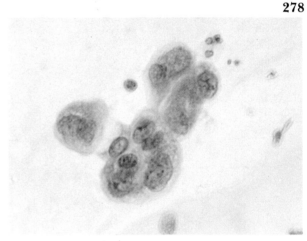

278

277 Adenocarcinoma. This is another example of dyskaryotic nuclei present in cells which form a vacuolated acinus. The patient was later diagnosed on curettage as having adenocarcinoma of the endometrium. (× 250)

278 Atypical hyperplasia of endometrium. Comparison with this field shows very similar nuclear abnormality. Comparability of cell morphology between well-differentiated adenocarcinoma and atypical hyperplasia is less surprising, as these can be difficult to distinguish on histological section of curettings. (× 250)

Examples of adenocarcinoma of endometrium

Case 1

279

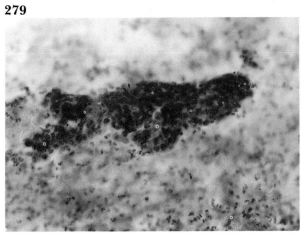

279 Papillary fragment. In this field malignant cells exfoliate in the form of a papillary fragment. (× *80*)

280

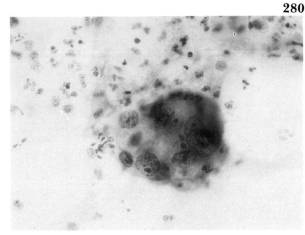

280 Acinar fragment: level 1. This is an acinar fragment found elsewhere. It has been photographed at three levels to demonstrate that it is three-dimensional. (× *160*)

281

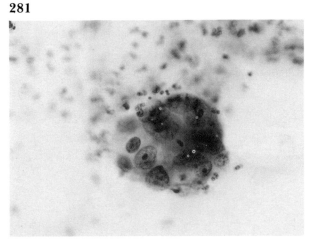

281 Acinar fragment: level 2. (× *160*)

282

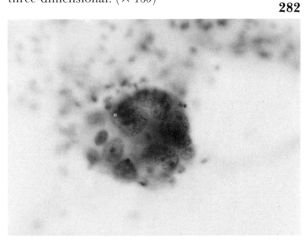

282 Acinar fragment: level 3. (× *160*)

283

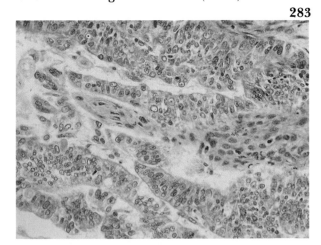

283 Acinar carcinoma of endometrium: hysterectomy. A section from the endometrium shows a mixed acinar and papillary pattern. (*H&E*, × *80*)

Case 2: Adenocarcinoma of endometrium

284

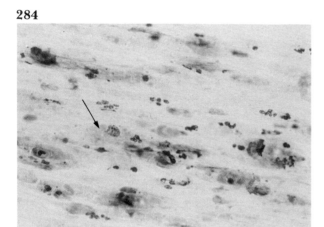

284 Degenerate tumour cells. This field shows amorphous debris, inflammatory cells and red cells with degenerate tumour cells which have plentiful foamy cytoplasm (arrow). (× *160*)

285

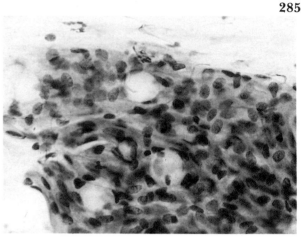

285 Surface endometrium. Elsewhere in the smear sheets of undifferentiated cells were found. Note that in this sheet of cells the nuclear changes are no more than dyskaryotic. Also note the gland openings. (× *160*)

286

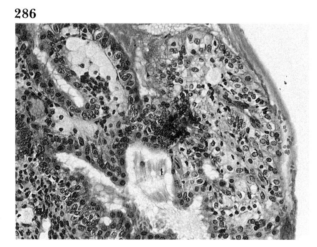

286 Adenocarcinoma of endometrium: curettings. Note the degeneration of the surface of this endometrial fragment, and the presence of foam cells. (*H&E,* × *80*)

287

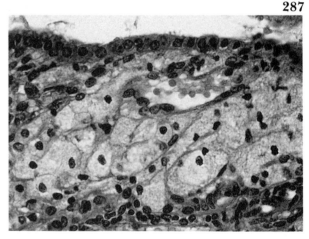

287 Adenocarcinoma of endometrium: curettings. This field has been taken at the same magnification as the cytology shown in **284** so that the morphology of the foam cells can be compared. These cells are reactive rather than neoplastic. (*H&E,* × *160*)

Case 3: Adenocarcinoma of endometrium

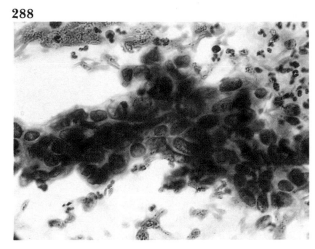

288 Tissue fragment. This field shows a typical tissue fragment shed from an adenocarcinoma. Note the community border at one edge and the suggestion of an acinar structure. In addition, the nuclei in the fragment have a three-dimensional appearance. (× *160*)

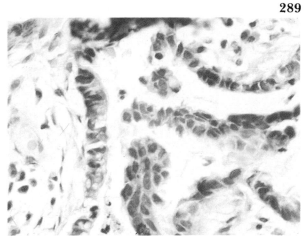

289 Adenocarcinoma of endometrium: curettings. Section of the curettings shows a well-differentiated adenocarcinoma. (*H&E, × 160*)

Case 4: Undifferentiated adenocarcinoma

290 Undifferentiated malignant cells. A cluster of undifferentiated malignant cells is seen in this field. There is nothing to indicate the type of differentiation or the primary site (endometrium or cervix). (× *160*)

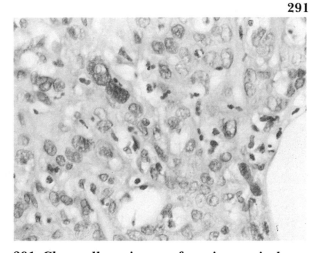

291 Clear cell carcinoma of cervix: cervical biopsy. This case is included to illustrate that there can be difficulties in distinguishing between tumours of endometrium and cervix. The cervical smear was reported with a preference for an endometrial tumour, but biopsy and fractional curettage showed a poorly differentiated clear cell adenocarcinoma of cervix. (*H&E, × 160*)

Case 5: Adenocarcinoma of endometrium

292

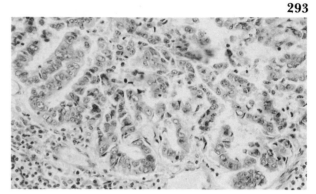

292 Poorly differentiated malignant cells. This is another example of poorly differentiated malignant cells, but fractional curettage and later hysterectomy showed this to be an adenocarcinoma of endometrium. (× *160*)

293

293 Adenocarcinoma of endometrium: hysterectomy. It is interesting that section of the endometrium shows a moderately well-differentiated adenocarcinoma. (*H&E*, × *80*)

Adenosquamous carcinoma and adenoacanthoma of endometrium

It is not uncommon to have benign squamoid change in an adenocarcinoma of endometrium, and this condition is called adenoacanthoma. More rarely the squamous component is also malignant, and the lesion is then an adenosquamous carcinoma. The distinction is important because of the less favourable prognosis in cases of adenosquamous carcinoma. The diagnosis is usually made histologically, although the probability of a squamous component can be predicted from the presence of abnormal squamous cells in the smear. However, it is not usually possible to decide whether such squamous cells are benign or malignant. Cases 6 and 7 illustrate this.

Case 6: Adenosquamous carcinoma of endometrium

294

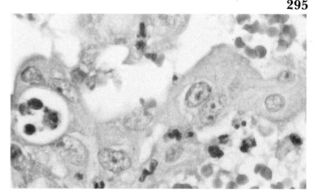

294 Malignant squamous and columnar cells. Other fields from this case were illustrated in **273** and **275**. In addition to malignant columnar cells, note the larger malignant cell which shows squamous differentiation of the cytoplasm. (× *250*)

295

295 Adenosquamous carcinoma of endometrium: curettings. Section of the endometrium shows the squamous component of the carcinoma with poorly differentiated malignant columnar cells to one side of the picture. (*H&E*, × *250*)

Case 7: Adenoacanthoma of endometrium

296 Malignant columnar cells. This field contains a cluster of malignant columnar cells with prominent nucleoli. There is also a single cell which shows squamous differentiation of the cytoplasm; this also shows nuclear chromatin irregularities and a prominent nucleolus. (× *400*)

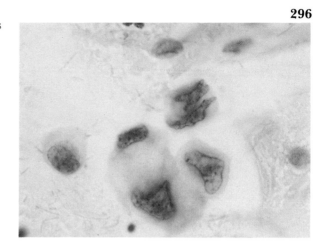

296

297 Dyskaryotic squamous cells. At lower magnification a cluster of abnormal squamous cells is seen, but the nuclear changes are no more than dyskaryotic. (× *160*)

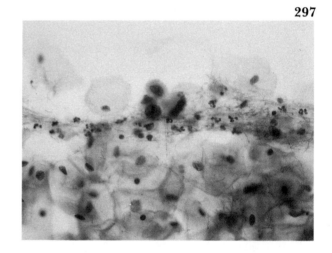

297

298 Adenoacanthoma of endometrium: curettings. Section of the endometrium shows an adenocarcinoma with benign squamoid change. (*H&E*, × *160*)

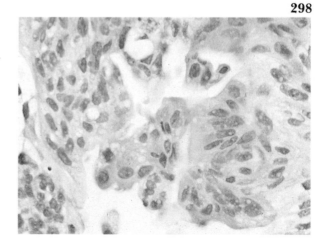

298

Adenocarcinoma of endometrium missed on curettage

It is important to remember that adenocarcinoma of endometrium can be missed on curettage (*Butler et al., 1971*). Because of this, due emphasis must be given to the presence of malignant columnar cells in the smear, and a negative curettage should not exclude further investigation.

Case 8

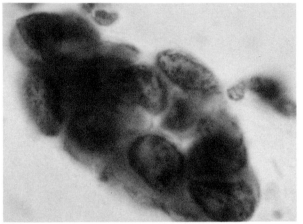

299 Malignant columnar cells. This field illustrates one of several clusters of cells in this case which were considered to be diagnostic of adenocarcinoma. (× *620*)

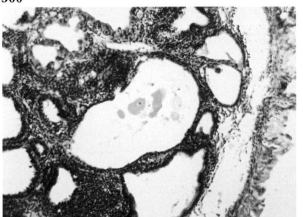

300 Atrophic endometrium: curettings. Curettage showed only dense stroma with dilated glands. This could be postmenopausal endometrium or part of a polyp. Note the reactive strip of surface columnar epithelium to one side of the field. (*H&E*, × *37.5*)

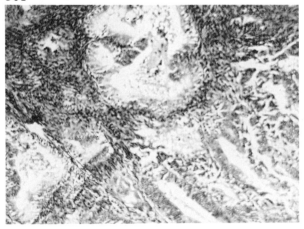

301 Adenocarcinoma of endometrium: hysterectomy. In this case it was fortunate that, for clinical reasons and in view of the cytology report, the gynaecologist proceeded to hysterectomy. Section of the endometrium showed a focus of adenocarcinoma at one cornu of the uterus. (*H&E*, × *37.5*)

Miscellaneous cases

Hydatidiform mole

302

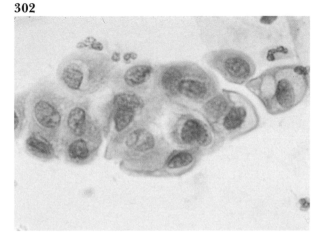

303

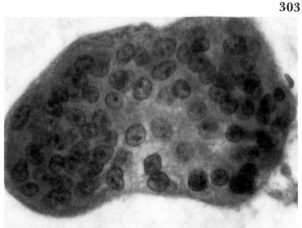

302 Part of a chorionic villus. This unusual tissue fragment was found in the cervical smear taken routinely from a pregnant woman who was suspected of having a hydatidiform mole. Note the brush border. (× *250*)

303 Chorionic villus: contact smear from mole. Following removal of the mole, contact smears were taken from the surface. Note the similarities of the cells, allowing for degeneration in the cervical smear specimen. In particular, note the brush border. (× *250*)

Cytology of endometrial aspirates

The difficulties in diagnosing endometrial lesions from cells present in cervicovaginal material have been illustrated. It might be expected that results would be more reliable with better-preserved material aspirated directly from the cavity of the uterus. Excellent results have been obtained by some (*Morse et al., 1982; Skaarland, 1986*), but interpretation is still difficult and, in addition to cytological expertise, considerable dexterity is necessary in the collection of the aspirate. Various methods have been introduced, but these are all based on direct aspiration through a cannula, washing under negative pressure and scraping the endometrium with a brush or microcurette. In some papers combinations of these methods are described. The material can be prepared as direct smears, on millipore filters or by section of cell blocks. Some authors advise the use of more than one preparation method.

In this section **304–311** were photographed, with kind permission of Dr Elsa Skaarland, from her own material. This was collected using the Gynoscann instrument (called Endoscann before 1983). Direct smears were made and fixed using an aerosol fixative, and staining was by Papanicolaou's method. The legends follow Dr Skaarland's diagnostic criteria and have been checked by her. Evaluation of the smears depends on a study of the architecture of the larger tissue fragments and the morphology of cells which present singly or in small fragments or clusters. The appearance of the background is also taken into consideration.

304 Normal endometrium. This field shows a
large tissue fragment which consists of glandular
structures surrounded by stroma. Note the regular
uniformity of the constituent cells. (× *40*)

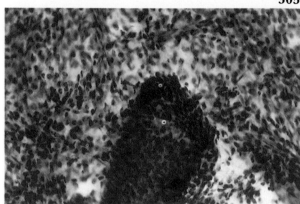

305 Normal endometrium. This field shows
stroma with capillaries running through it, and a
superimposed glandular fragment. (× *40*)

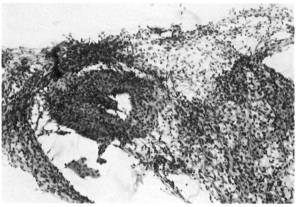

In **304–306** the background shows no evidence
of necrosis.

306 Normal endometrium. This example shows
surface epithelium with a gland opening. (× *40*)

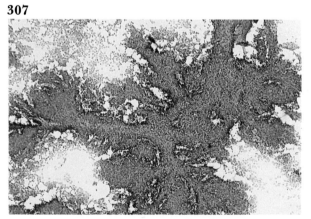

307 Adenomatous hyperplasia. This fragment
consists of close-packed glandular structures with no
stroma. The background consists of blood only.
(× *25*)

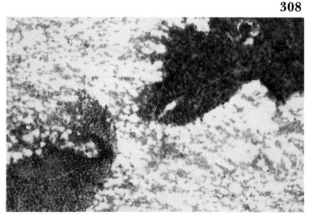

308 Adenocarcinoma of endometrium. This
low-power field contrasts a fragment of normal
endometrium (which consists of surface epithelium
and stroma, with part of a gland opening) with a
sheet of larger malignant cells showing a loss of
regular growth pattern. (× *40*)

309 Adenocarcinoma of endometrium. At higher magnification cell morphology can be studied. There is some pleomorphism, the nuclear chromatin pattern is coarse and irregular, and prominent nucleoli are seen. (× *160*)

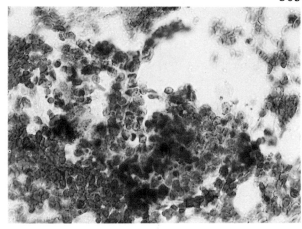

310 Adenocarcinoma of endometrium. Elsewhere in the same smear a loose cluster of disaggregated pleomorphic malignant cells is seen. Some cells look rounded (see **288**) and most have large, prominent nucleoli. Note the normal cells which are also present: in some cases of adenocarcinoma these can dominate the cell picture. (× *160*)

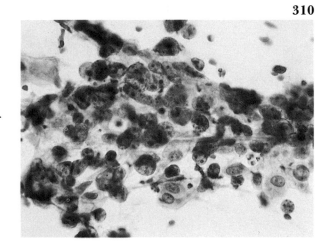

311 Adenocarcinoma of endometrium. This field is characteristic, in that it shows a colony of cells with a vascular core which is part of the ordinary connective tissue support. (× *160*)

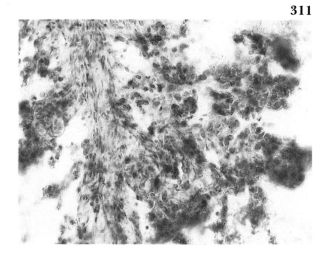

312

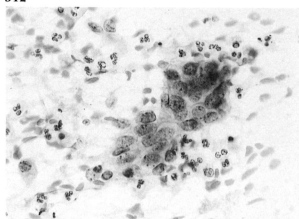

312 Acinar fragment. The endometrial aspirate in this illustrative case was collected by an Isaacs sampler. A well-preserved cluster of cells in the form of an acinar fragment is seen. There is a sero-sanguinous background with necrotic cells. The variation in size indicates that these are not trichomonads. (× *160*)

313

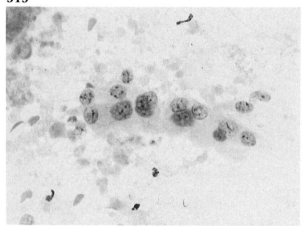

313 Malignant columnar cells. Another field in the same smear shows a strip of malignant columnar cells. (× *160*)

314

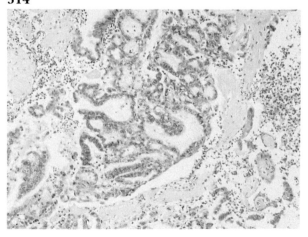

314 Adenocarcinoma of endometrium: curettings. Section of the endometrium shows a moderately well-differentiated adenocarcinoma with areas of necrosis. (*H&E*, × *40*)

Cell block method

As shown above, much of the material collected from the endometrial cavity is in the form of tissue fragments or microbiopsies. When concentrated into a cell block and processed histologically, sections are similar to sections of endometrial curettings. This has the advantage of showing tissue pattern, but individual cell morphology may be blurred. As the block consists of multiple fragments with a range of histological appearances, it is essential to make serial sections of the block examining every tenth section.

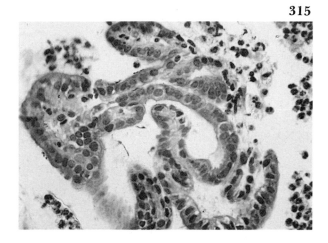

315 Normal gland: section of cell block. This section shows an inactive endometrial gland from a postmenopausal woman. There is no stroma, but focal collections of polymorphs are seen around the gland complex. (*H&E*, × *125*)

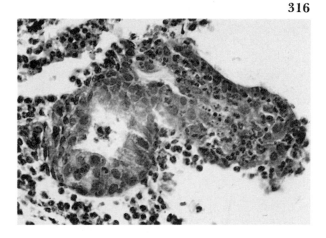

316 Hyperplasia: section of cell block. In this field the gland is lined by hyperplastic cells, but without a complete tissue pattern to recognize invasion there is not enough cytological irregularity to diagnose malignancy. (*H&E*, × *125*)

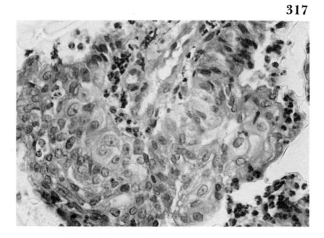

317 Adenocarcinoma: section of cell block. This example shows better cell preservation. The constituent cells are pleomorphic with nuclear irregularity and prominent nucleoli, which is consistent with a diagnosis of adenocarcinoma. The diagnosis was confirmed on hysterectomy. (*H&E*, × *125*)

8 Cytology of other sites

Vulva

The normal vulva is covered by skin which is the same as skin elsewhere, in that it matures to a keratinized layer. Because of this, smears consist of anucleate keratinized cells.

Vulval lesions include the vulval dystrophies, vulval intraepithelial neoplasia (VIN) and invasive squamous cell carcinoma. Adenocarcinoma of Bartholin's gland has been reported, and the vulva can be a primary site for malignant melanoma. Cytological diagnosis is of limited value because in most vulval disease the surface is covered by hyperkeratosis and parakeratosis, which prevents exfoliation of more severely abnormal cells from deeper layers. In these cases the smear can underestimate the severity of the lesion, and this would be expected in VIN III of the Bowenoid type, whereas when the VIN III is of the small basiloid type there is less parakeratosis and exfoliation should be more profuse (*Buckley et al., 1984*). Except in cases of overt invasive cancer, clinical presentation is similar in vulval dystrophies, VIN and micro-invasive cancer. The woman complains of marked irritation or pain, and on examination white patches are present. Colposcopic examination may show vascular areas between the white patches, and directed smears from these sites are more reliable, but in general the diagnosis of vulval disease must be by multiple biopsies.

Case 1: VIN III (Bowenoid)

318

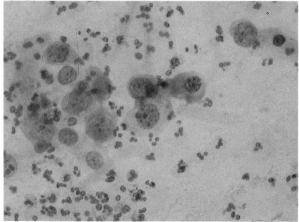

319

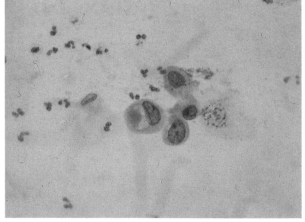

318 Reactive and dyskaryotic cells. This is an infected smear with reactive and dyskaryotic cells present; one of the latter is keratinized. Reactive changes are common because trauma caused by scratching causes regenerative cell changes. (× *160*)

319 Dyskaryotic cells. Elsewhere in the smear degenerate nuclei are seen, as well as cells with reactive and dyskaryotic nuclei. This case is unusual, in that there is reasonable exfoliation but biopsy showed VIN III of the Bowenoid type. (× *160*)

Case 2: Paget's disease of vulva

320

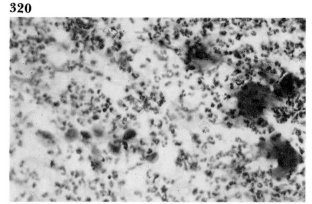

320 Infection and cell degeneration. In this case there is more marked infection of the smear, with cell degeneration. Note the keratinized cells. (× *160*)

321

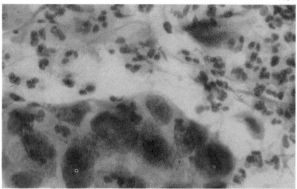

321 Severely dyskaryotic cells. At higher magnification a sheet of severely dyskaryotic cells is seen. These show pleomorphism, with angular nuclear irregularities and occasional prominent nucleoli. The cytological picture suggests the possibility of invasive cancer, but histology showed that this was a case of Paget's disease. (× *400*)

Case 3: VIN III (basiloid)

322

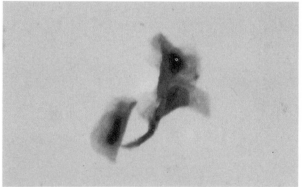

322 Keratinized dyskaryotic cells. Mildly dyskaryotic cells are seen in this smear. Note the blurred staining; this is common in smears collected from vulval lesions. (× *125*)

323

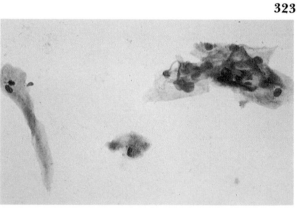

323 Mild dyskaryosis. In spite of being a case of basiloid VIN III, this was a scanty smear and this field shows the remaining dyskaryotic cells present. (× *125*)

324

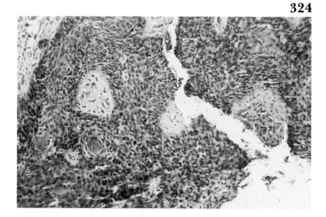

324 VIN III (basiloid): simple vulvectomy. Sections showed the lesion to be predominantly small cell in type, but with differentiation on the surface which prevented exfoliation of the more severely dyskaryotic cells. (*H&E*, × *37.5*)

Case 4: VIN III

325

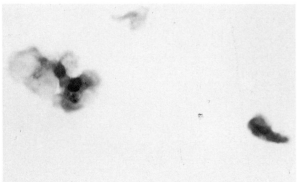

326

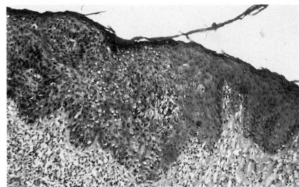

325 Degenerate dyskaryotic cells. This is another example of a very scanty smear. The cells present were mainly anucleate keratinized squames with a few degenerate, mildly dyskaryotic cells, as seen in this field. (× *125*)

326 VIN III: simple vulvectomy. In this section a thick layer of hyperkeratosis prevents exfoliation of the severely dyskaryotic cells in the deeper layers. Note the abrupt transition between normal and abnormal epithelium in this section. (*H&E*, × *37.5*)

Case 5: Squamous carcinoma of vulva— early invasion

327

328

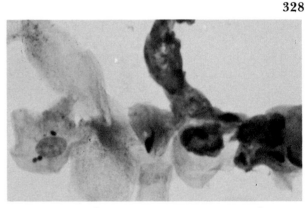

327 Portrait of vulva. White patches are seen on the labia minora, and on clinical inspection this woman could have vulval dystrophy, VIN or early invasive cancer.

328 Mild dyskaryosis and cell degeneration. The smear was scanty and this field is representative. It shows keratinized anucleate squames, with a mildly dyskaryotic cell and a cluster of degenerate cells. (× *250*)

329

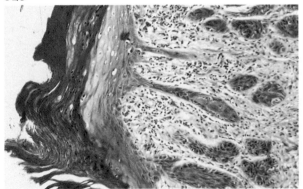

329 Squamous carcinoma of vulva: early invasion— simple vulvectomy. Section shows very dense hyperkeratosis and, deep to this, squamous carcinoma with early invasion of the stroma. (*H&E*, × *37.5*)

Case 6: Invasive squamous cell carcinoma (recurrence)

330 Poorly differentiated malignant cells. This patient had already had a squamous cell carcinoma of vulva treated by simple vulvectomy. This smear was taken from a vascular nodule that was clinically diagnosed as recurrent carcinoma. This field shows keratinized cells with a few poorly differentiated malignant cells. (× 125)

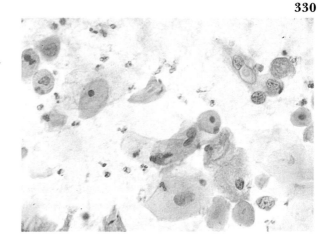

330

331 Malignant pearl. This field contains a keratinized malignant pearl. (× 250)

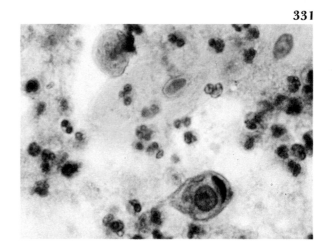

331

332 Malignant pearl. Elsewhere in the smear a second pearl was found, with a cluster of degenerate cells. The diagnosis of recurrent carcinoma was confirmed on biopsy. (× 250)

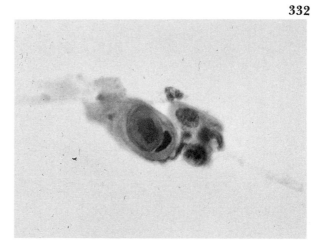

332

Case 7: Squamous cell carcinoma of clitoris

333

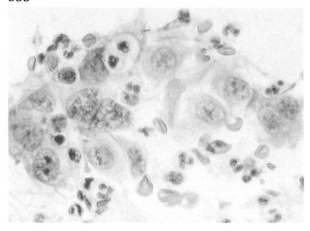

333 Malignant squamous cells. This case is an example of squamous cell carcinoma in which malignant cells exfoliate freely. This field shows a sheet of well-preserved malignant squamous cells. (× 250)

334

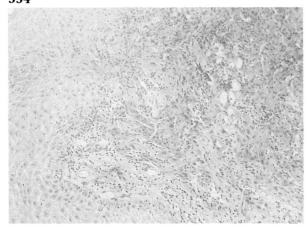

334 Squamous cell carcinoma: biopsy of tumour. At low magnification the surface squamous epithelium is seen, with tongues of tumour infiltrating into the stroma. (H&E, × 37.5)

335

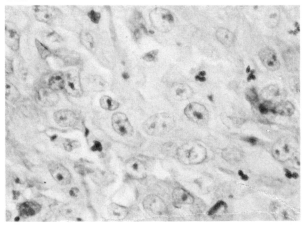

335 Squamous cell carcinoma: biopsy of tumour. Part of the section taken at the same magnification as the vulval smear makes it possible to compare cell morphology. (H&E, × 250)

Case 8: Malignant melanoma of vulva—amelanotic

336

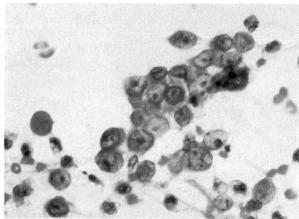

337

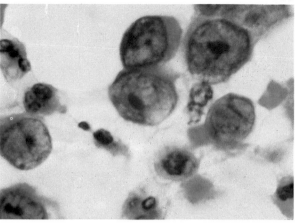

336 Malignant cells. This woman also had a tumour of clitoris, which was similar to that in the previous case on clinical inspection. However, the cytological picture is quite different. This field shows undifferentiated cells with macronucleoli. (× 250)

337 Malignant cells. At higher magnification a notch is seen clearly in one nucleus and the macronucleoli show sharp angularities. In spite of the absence of pigment, the cell morphology warranted a diagnosis of malignant melanoma. (× 620)

338

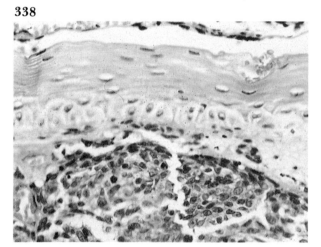

339

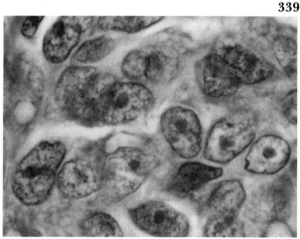

338 Malignant melanoma of vulva: biopsy. This section includes surface epithelium and shows junctional activity but minimal pigment. (H&E, × 125)

339 Malignant melanoma of vulva: biopsy. At the same magnification as the cells in **337** the morphology can be compared. Notching of the nuclear membrane and irregularities of the nucleoli are more obvious in the tissue. (H&E, × 620)

Vagina

A case of benign vaginal adenosis is illustrated in **108–112**. That patient had not been exposed to DES in utero, so this was an example of congenital vaginal adenosis. Cases of adenocarcinoma developing in DES-exposed girls have been rarer in this country than in the USA, but the case presented here is one of them (*Buckley et al., 1982*). The patient had been exposed to DES in the first trimester of her mother's pregnancy. She had an early menarche but was otherwise well until she began to take an oral contraceptive at the age of 16 years. By 17 years of age she presented with post-coital bleeding, and tumour was found on clinical examination.

Case 1: Vaginal adenocarcinoma—clear cell

340

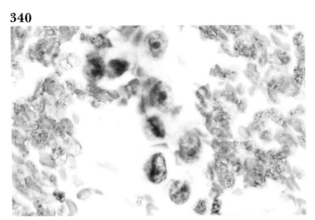

340 Undifferentiated malignant cells. The smear was bloodstained with an almost pure population of undifferentiated malignant cells. These show pale-staining clear cytoplasm, and the nuclei contain prominent red nucleoli. (× 250)

341

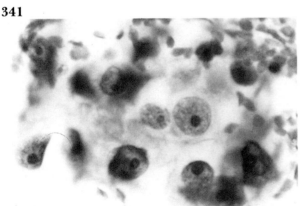

341 Undifferentiated malignant cells. The clear cytoplasm is more apparent in this sheet of cells found in another field. (× 250)

342

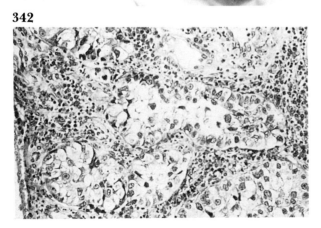

342 Vaginal adenocarcinoma (clear cell): colposcopic biopsy. Treatment was by extended hysterectomy and vaginectomy, and examination of the specimen showed that the tumour arose in the vaginal fornices. The cervix was clear of tumour with the exception of spread from the fornices to the periphery of the ectocervix. The histological picture of the tumour was similar to that seen in this section from a colposcopic biopsy. This shows a poorly differentiated clear cell adenocarcinoma. (*H&E*, × 62)

Ovary

The cytology of ovarian tumours is usually dealt with in studies of peritoneal fluid or fine-needle aspiration of ovarian masses, and readers are referred to texts in which these are discussed for further details. This section illustrates the occasional cases in which examination of vaginal or cervical smear suggests the probable presence of ovarian carcinoma.

Tissue fragments from ovarian cancers can pass through the fallopian tube and uterine cavity to appear in the cervical smear without spread to endometrium being present. The characteristic features are the finding of tumour fragments and single cells in a clean smear with no cancer diathesis. The fragments are often moulded by the fallopian tube to have a cast-like appearance. In some cases psammoma bodies are present, but these are not diagnostic as they can be seen in endometrial cancers and in reaction to an IUCD (see **88** and **89**). In older women tumour fragments may be accompanied by an unexpected oestrogen effect, but sheets of parabasal cells as seen in an atrophic smear, without vaginitis, are also common. It is seldom possible to identify the tumour type because of cell degeneration and blurred staining.

Case 1: Serous cystadenocarcinoma of ovary

343

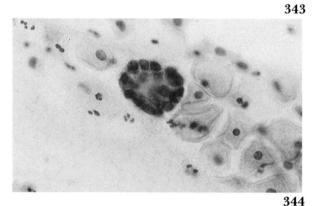

343 Papillary fragment: vaginal smear. This field contains a papillary fragment, seen en fosse, against a background of parabasal and small intermediate cells. (× *160*)

344

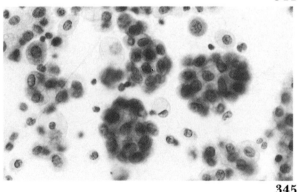

344 Pleural fluid. This patient had a pleural effusion, and this field from a fluid preparation is shown for completeness. The appearances are similar, but cells and fragments have rounded up after being suspended in fluid. (× *160*)

345

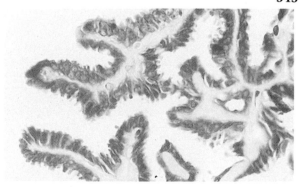

345 Serous cystadenocarcinoma of ovary: ovary. Following surgery the only tumour found was in the ovary, and this section shows a well-differentiated papillary cystadenocarcinoma. (*H&E*, × *160*)

Case 2: Mucinous adenocarcinoma of ovary

346

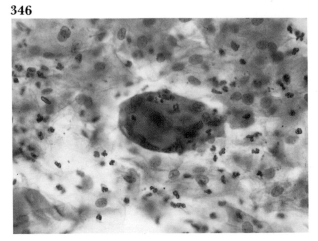

347

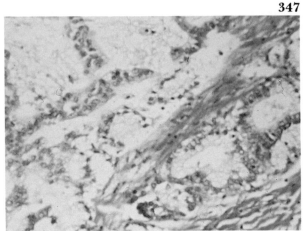

346 Degenerate tumour fragment. In this case the background is similar, but the tumour fragment shows more degeneration and has been compressed to give it a cast-like form. (× *125*)

347 Mucinous adenocarcinoma: ovary. Section shows a poorly differentiated mucinous adenocarcinoma with tissue necrosis. (*H&E, × 62*)

Case 3: Adenocarcinoma of ovary

348

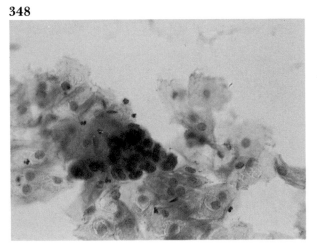

349

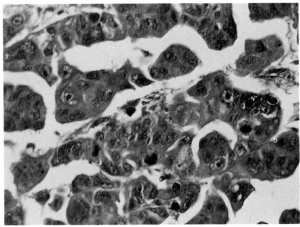

348 Tumour fragment. This fragment was an incidental finding in a routine cervical smear taken from a woman aged 48 years. The fragment is seen against a background of normal intermediate cells. (× *250*)

349 Adenocarcinoma of ovary: ovary. Further investigation showed no abnormality of cervix or endometrium, and an ovarian tumour was found at laparotomy. This field, at the same magnification as the cervical smear, shows a poorly differentiated adenocarcinoma. (*H&E, × 250*)

Case 4: Adenocarcinoma of ovary

350 Malignant cells. This example differs from those already illustrated, in that the background is serous and the smear lacks the 'clean' appearance of the other cases. In addition, the malignant cells form a looser cluster without the compression previously seen. Uterine spread might be suspected in this case. (× *120*)

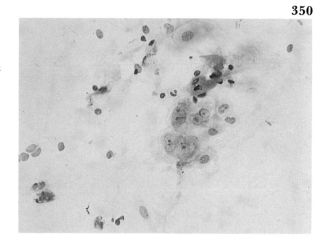

351 Adenocarcinoma of ovary: ovary. Section of the ovary shows a poorly differentiated adenocarcinoma with tissue necrosis. There was no evidence of spread to the endometrium. (*H&E*, × *66*)

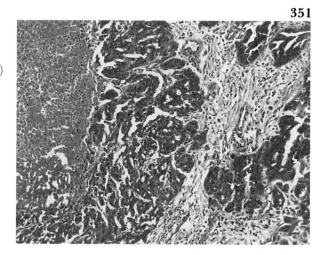

352 Adenocarcinoma of ovary: ovary. At higher magnification the cells in the tissue can be compared with the cells in the smear. (*H&E*, × *132*)

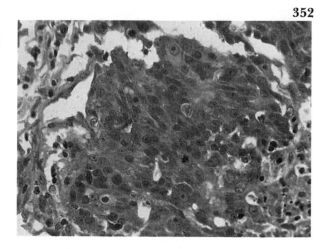

Case 5: Follicular cyst

This case is included for interest. Occasionally gynaecologists find large follicular cysts at laparoscopy or laparotomy and send the aspirated fluid for cytology. As the granulosa cells lining such cysts are immature, the appearances can be alarming if they have not been seen before.

353

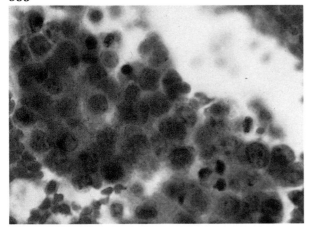

353 Fluid from a follicular cyst (courtesy Dr Hilda Harris). Sheets of small cells with granular nuclei and blurred margins are present, and in this field at least one mitosis can be identified. (× 250)

354

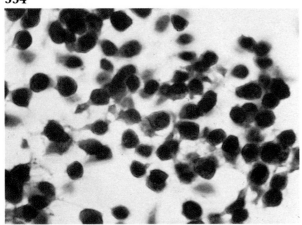

354 Comparison: exfoliated cells in a section of a follicular cyst. This field is taken at the same magnification as the fluid preparation. The cells look smaller because of the histological method of preparation, but similarities can be recognized. (H&E, × 250)

355

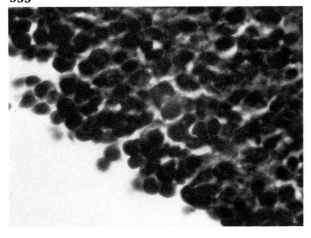

355 Comparison: section of granulosa cell layer in a follicular cyst. The cells in the tissue can be compared with those exfoliated into the cyst and with those in the cytological preparation. Note the blurred outline. (H&E, × 250)

9 Problem cases

In previous chapters an attempt has been made to illustrate the range of cytological appearances found both with specified cell types and also with different pathological lesions. This chapter presents cases at greater length to illustrate various problem areas in gynaecological cytology. These include the need to recognize multiple types of abnormal cells in combined lesions, the apparent failure of cytological–histological correlation and differential diagnosis in undifferentiated small-cell lesions.

Combined lesions

Two cases which fall into this category are illustrated in the 'miscellaneous' section of Chapter 6.

Case 1: Adenocarcinoma of cervix with CIN III

The cervical smear in this case contains well-differentiated squamous dyskaryotic cells with dyskaryotic and malignant columnar cells. Recognition of malignant columnar cells is important, as dyskaryotic columnar cells can overlie a CIN lesion. In this case they could come from the periphery of the adenocarcinoma.

356

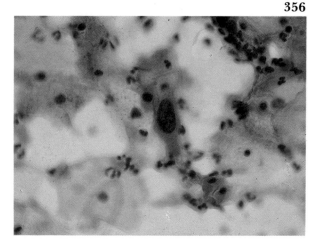

356 Dyskaryotic squamous cells. This field shows a keratinized, severely dyskaryotic squamous cell with some nuclear degeneration. (× *160*)

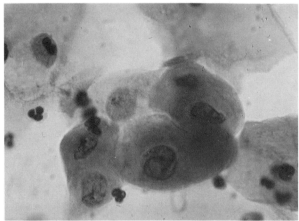

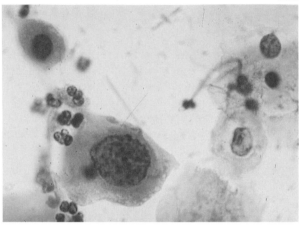

357 Dyskaryotic squamous cells. Elsewhere in the smear, at higher magnification, nuclei in this cluster of dyskaryotic cells show hydropic degeneration. There is also degenerative condensation of chromatin, particularly under the nuclear membrane. (× *400*)

358 Dyskaryotic squamous cell. This single dyskaryotic cell shows coarse granularity of the chromatin, with degenerative wrinkling of the nuclear membrane. In spite of the presence of small vacuoles in the cytoplasm, the cytoplasmic border is sharply demarcated. (× *400*)

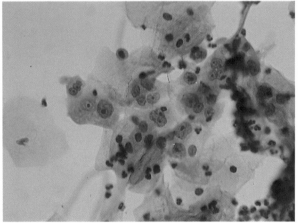

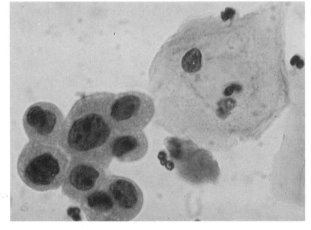

359 Dyskaryotic glandular cells. This group of dyskaryotic glandular cells shows pleomorphism with eccentric nuclei, floccular cytoplasm and prominent nucleoli. (× *160*)

360 Dyskaryotic glandular cells. At first glance this group, at higher magnification, could be squamous metaplastic cells as some of the nuclei are placed centrally, which suggests that these are squamous cells. However, the single centrally placed, prominent, red nucleoli point to a glandular origin. The nuclear chromatin pattern is uniformly granular, so the cells cannot be called more than dyskaryotic (pre-invasive). Note also the floccular cytoplasm in keeping with glandular cells. (× *400*)

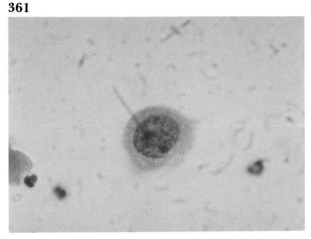

361 Single dyskaryotic glandular cell. This single cell is similar, but there is a suggestion of clearing in the nuclear chromatin pattern. (× *400*)

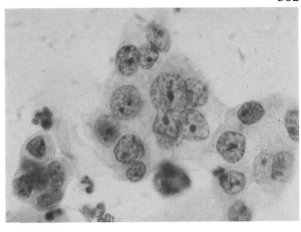

362 Malignant glandular cells. Elsewhere in the smear, clusters of glandular cells are found showing criteria of malignancy. Note the blurring of the cytoplasmic borders and irregular condensation of chromatin, both in the nucleus and at the nuclear membrane. As before, most of the nucleoli are single and centrally placed, but they vary in size and shape. (× *400*)

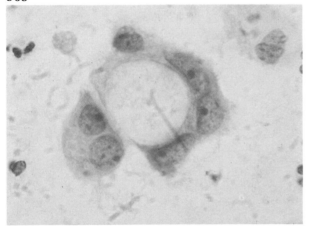

363 Malignant glandular cells. The diagnosis of adenocarcinoma is confirmed by this tubular structure found elsewhere in the smear. (× *400*)

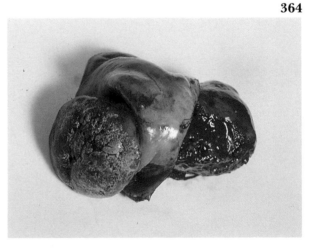

364 Hysterectomy specimen. The body of the uterus is seen to the right of the picture and the posterior lip of the cervix is replaced by a tumour with surface necrosis. This was the adenocarcinoma, and the smooth anterior lip, which lies above, was found to have extensive CIN III of the well-differentiated keratinizing type.

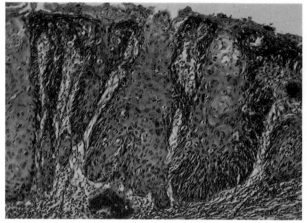

365 CIN III (keratinizing): anterior lip of cervix. This section illustrates pleomorphism and a disordered growth pattern throughout the thickness of the epithelium. The dermal papillae are irregularly placed and, although the anterior lip of the cervix was normal on naked-eye examination, an abnormal vascular pattern would have been seen colposcopically if this investigation had been available when this case was diagnosed and treated. (*H&E, × 40*)

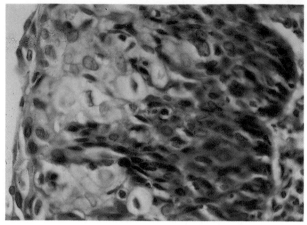

366 CIN III (keratinizing): anterior lip of cervix. This section, at higher magnification, allows comparison of the surface cells with those seen in the cervical smear. Note the occasional koilocytes in the deeper layers of the tissue. (*H&E, × 160*)

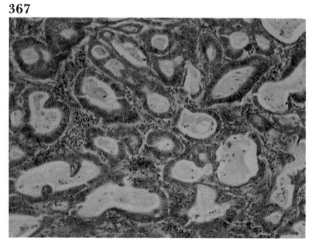

367 Adenocarcinoma: posterior lip of cervix. Sections taken from the tumour replacing the posterior lip of the cervix show a moderately well-differentiated adenocarcinoma of the endocervical type. (*H&E, × 40*)

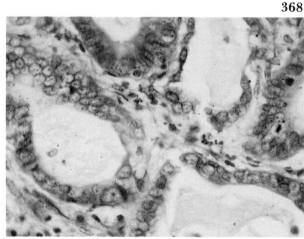

368 Adenocarcinoma: posterior lip of cervix. At higher magnification it is possible to compare the cellular morphology with that of the malignant glandular cells seen in the cervical smear. Note the similarities of the nuclear chromatin pattern and that of the nucleoli. (*H&E, × 160*)

Discrepancies of cytological–histological correlation

The cases presented in this section illustrate some of the circumstances in which this can occur. Underestimation of the degree of abnormality in cervical smears occurs when smears are suboptimal and when relatively few abnormal cells are present. Meticulous re-screening in these cases usually identifies a few cells which should have caused suspicion of abnormality. This is the type of case which is missed when cytology screeners are under pressure to screen excessive numbers of slides, and demonstrates the need for adequate staffing in cytology laboratories (*Report of the Intercollegiate Working Party on Cervical Cytology, 1987*). In other cases the appearances may be misinterpreted, and this usually results from infection or degeneration. There are also cases when the cytological prediction of abnormality is not confirmed by cervical biopsy (*Byrne et al., 1988*). With the increased incidence of HPV infection, exfoliation may be prevented by hyperkeratosis or parakeratosis, and the presence of keratinized cells in the smear is a marker to indicate the need for careful follow-up.

Cases 2 to 7 are taken from a study in which 902 women were screened colposcopically as well as by taking cervical smears. Twenty-nine women showed colposcopic abnormality while their cervical smears were negative, or were reported as having inflammatory or borderline changes only. Review of histological and cytological material was possible in 11 cases, providing 22 cervical smears and 14 biopsy specimens. An attempt was made to identify markers in the cervical smears which might alert cytologists to the need for further follow-up. In parallel, a control series of 22 known negative smears was studied applying the same criteria. The most striking features were as follows.

1 The number of smears considered to be suboptimal was doubled.

2 In 25 per cent of the cervical smears dyskaryotic and/or koilocytes were identified on review.

3 Keratinized cells, singly and in sheets, were more common in the study group than in the control group and dense sheets of cells (see **382**) were also seen more often.

4 Review of the biopsy material modified the original histological diagnosis in 50 per cent of the cases.

Case 1: CIN III—missed in the first smear

369

370

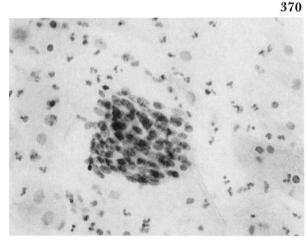

369 Undifferentiated dyskaryotic cells. This infected smear was taken from a woman who had had a negative smear less than a year previously. A number of sheets of undifferentiated dyskaryotic cells were present and one is seen in this field; this could be squamous or glandular in origin. (× *80*)

370 Undifferentiated dyskaryotic cells. In another sheet, at higher magnification, nuclei show a granular chromatin pattern and some nuclei are fusiform in shape. (× *160*)

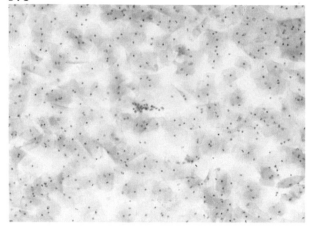

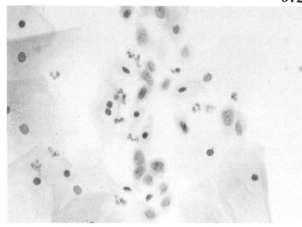

371 Missed dyskaryotic cells. The negative smear was reviewed and, at the usual screening magnification, a mature cell pattern was seen with occasional groups of small cells with hyperchromatic nuclei. (× *40*)

372 Missed dyskaryotic cells. Although rather pale, at higher magnification a similar group of cells can be recognized as severely dyskaryotic. (× *160*)

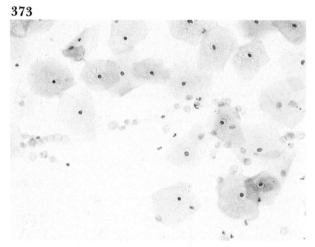

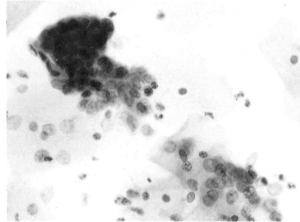

373 Large, pale, bare nuclei. Elsewhere in the smear collections of pale, bare nuclei were found. These may have sinister implications and should prompt the observer to screen more carefully or request a repeat smear for further elucidation. (× *80*)

374 Dyskaryotic columnar cells. This tissue fragment of endocervical columnar cells with underlying 'reserve cells' might also have caused suspicion of abnormality. Cone biopsy confirmed the presence of CIN III. (× *160*)

Case 2: CIN III—suboptimal smear

375 Dyskaryotic parabasal cell. The first smear taken from this patient was scanty and unreliable. This is a field from the second smear which was reported as containing occasional dyskaryotic cells. A single dyskaryotic parabasal cell is seen in this field, and even when only a few such cells are found the probable presence of at least CIN II is indicated. (× *160*)

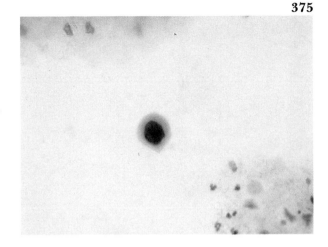

376 Dyskaryotic cells. On review, these moderate to severely dyskaryotic cells were found. (× *160*)

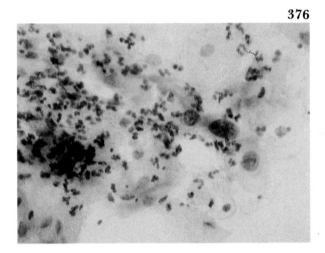

377 CIN III with koilocytosis: cone biopsy. The colposcopic biopsy showed features causing suspicion of microinvasion, so the patient was treated by cone biopsy rather than laser. Sections from the cone biopsy showed CIN III but no evidence of invasion. Note the koilocytes in the superficial layers: it is interesting that these did not exfoliate to present in the cervical smear. (*H&E*, × *40*)

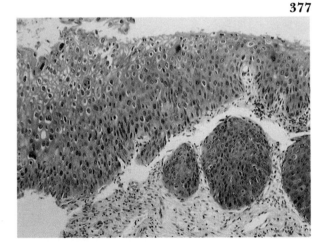

378

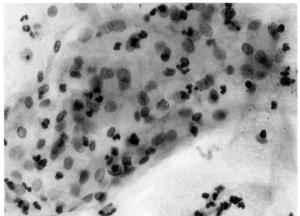

378 Metaplastic cells. In this case infected smears obscured diagnosis. The woman was followed up at 6-monthly intervals because of her colposcopic abnormality. This field is taken from the second smear, and shows a sheet of metaplastic cells with blurred nuclei. The changes do not amount to dyskaryosis. (× *160*)

379

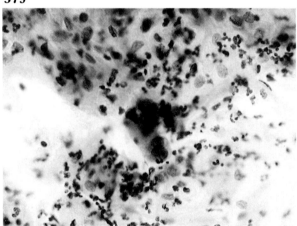

379 Degenerate cells. This field is taken from the third smear, and it contains a dense fragment which consists of degenerate cells with hyperchromatic nuclei which are difficult to evaluate. (× *160*)

380

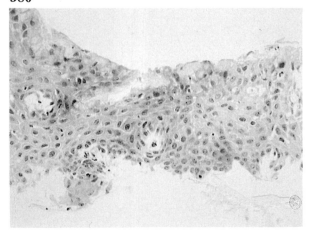

380 CIN I/II: colposcopic biopsy. This small biopsy shows atypical reserve cells deep to columnar epithelium. The original diagnosis was CIN II, but this was changed to CIN I/II on review. The patient received laser treatment. (*H&E*, × *80*)

Case 4: CIN (uncertain grade) with HPV infection

381

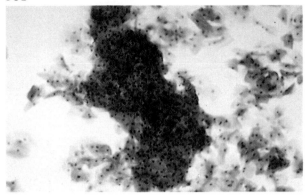

381 Keratinized cells. The presence of keratinized cells was noted in the original report. This field shows one of a number of keratinized sheets which were present. (× *40*)

382

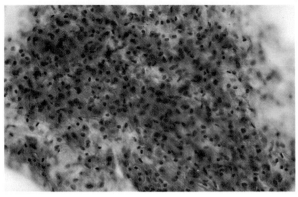

382 Dense sheets of cells. This is an example of a dense sheet of cells (see the introduction to this section). (× *80*)

383

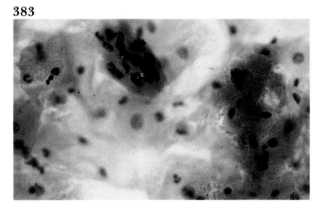

383 Degenerate keratinized cells. Another field from this cervical smear contains keratinized cells with degenerate granular nuclei. (× *160*)

384

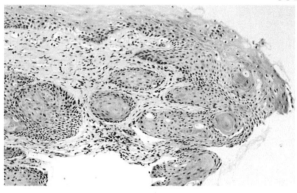

384 CIN (uncertain grade): colposcopic biopsy. The colposcopic biopsy included a focus of abnormal epithelium. As the tissue is cut tangentially, this could be no more than basal cell hyperplasia, but it could also be CIN of uncertain grade. Note the epithelial pearl present, and the koilocytes. (*H&E, × 40*)

385

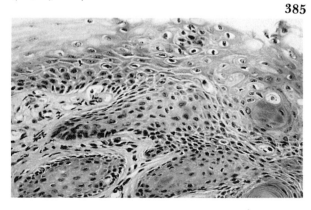

385 CIN (uncertain grade): colposcopic biopsy. At higher magnification the dyskaryotic nuclei in the tissue can be recognized. Treatment was by laser. (*H&E, × 80*)

Case 5: CIN I/II—suboptimal smears

386

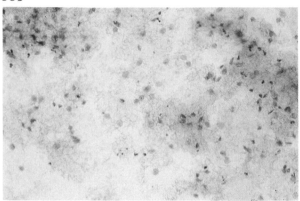

387

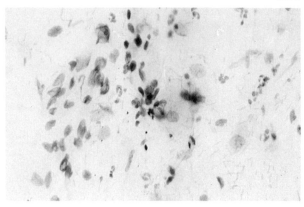

386 Cytolysis. This is another example of a woman who had a series of suboptimal smears. The field illustrated is from the second smear, and shows marked cytolysis. (× *80*)

387 Degenerate dyskaryotic cells. This field is taken from the third smear, which was reported as negative. However, review showed poor staining, with cell degeneration and with a number of groups of degenerate, hyperchromatic bare nuclei as seen in this picture. (× *160*)

388

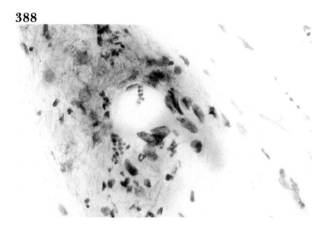

389

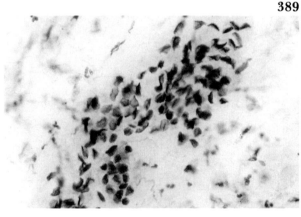

388 Degenerate keratinized cells. Other cells in the smear with nuclear degeneration showed keratinization. (× *160*)

389 Degenerate dyskaryotic columnar cells. Although precise diagnosis is not possible when there is cell degeneration, the presence of cells of the type illustrated here and in the previous fields (**387** and **388**) makes further follow-up mandatory. (× *160*)

390

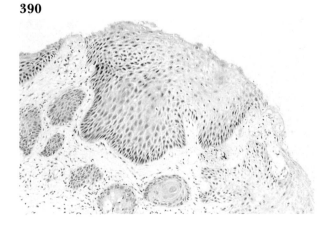

390 CIN I/II with keratinization: colposcopic biopsy. The biopsy was reported as CIN I/II. Note the keratinization on the surface. (*H&E, × 40*)

Case 6: Basal cell hyperplasia

391

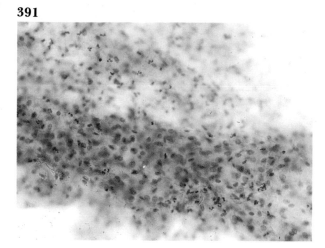

392

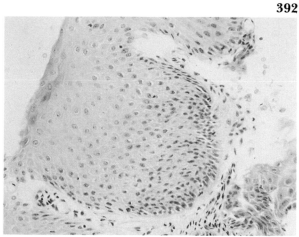

391 Dense sheet of cells. This patient also had a colposcopic abnormality. The cervical smear was reported as negative, and on review the only possible abnormality was the dense sheet of cells seen in this field. (× 80)

392 Basal cell hyperplasia: colposcopic biopsy. This section was originally reported as CIN I, but on review it was agreed that it was no more than basal cell hyperplasia. Note the keratinization of the surface which would account for the colposcopic abnormality. Further smears continued to be negative. (H&E, × 80)

Case 7: CIN III—missed on first biopsy

393

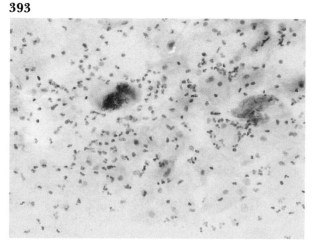

394

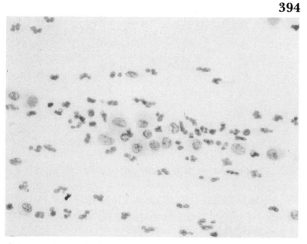

393 Mildly dyskaryotic cells. This smear was originally reported as showing inflammatory changes only, but on review multinucleated and mildly dyskaryotic cells were found, as illustrated in this field. (× 80)

394 Pale-staining dyskaryotic cells. Elsewhere in the smear, cells were found which were pale-staining but had a granular nuclear chromatin pattern (see **172**). (× 160)

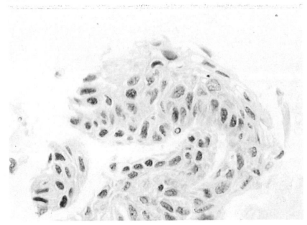

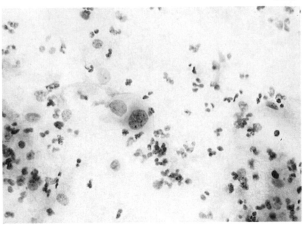

395 CIN uncertain grade (missed on first report): colposcopic biopsy. This colposcopic biopsy was originally reported as consisting of normal squamous epithelium only. Review showed the presence of the fragment of CIN illustrated in this field. (*H&E*, × *160*)

396 Dyskaryotic cells. Follow-up was continued because of the persistence of colposcopic abnormality, and a cervical smear taken 6 months after the first biopsy was reported as containing mildly dyskaryotic cells together with inflammatory changes and cell degeneration. These cells are seen in this field. (× *160*)

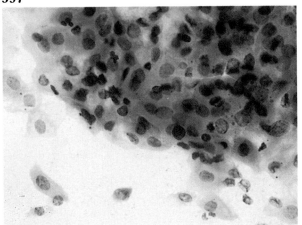

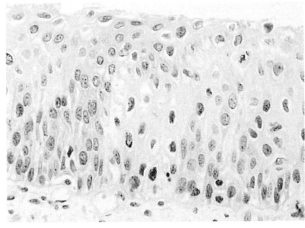

397 Dense sheet of dyskaryotic cells. Elsewhere in the smear, dense sheets of cells were found on review (see **382**). This illustration shows the periphery of one sheet in which cell morphology can be recognized and the cells can be seen to be moderately dyskaryotic. (× *160*)

398 CIN III: second colposcopic biopsy. The second biopsy was reported as showing CIN III. Because the woman also had uterine prolapse, treatment was by vaginal hysterectomy. Sections of the cervix were reported as showing foci of CIN III with HPV infection. (*H&E*, × *160*)

Case 8: Loss of epithelium

399

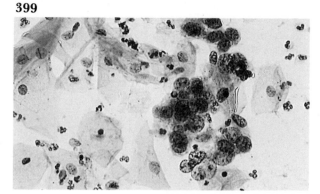

400

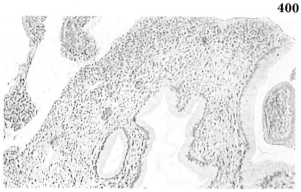

399 Severe dyskaryosis. This case is included to draw attention to the need for the careful handling of tissue, both in the theatre and in the laboratory. These poorly differentiated cells show a suggestion of clearing in the nuclear chromatin pattern, and prominent nucleoli are present. The smear was reported as showing features causing suspicion of an invasive lesion. (× *125*)

400 Loss of epithelium: cone biopsy. Sections of all blocks taken from the cone biopsy showed complete loss of covering epithelium, as seen in this field. This is an extreme example, but when severely abnormal cytology has been reported it is important that the histology report includes a comment on any loss of surface epithelium, particularly when the histology report does not support the cytological findings. (*H&E*, × *37.5*)

Undifferentiated small cells

The next group of cases illustrates the range of lesions which can present with small abnormal cells in the cervical smear. These may be found as single cells, when it is often possible to identify nuclear and cytoplasmic features which point to the nature of the lesion, but in many cases these cells are found in dense sheets or clusters which make assessment of individual cell morphology very difficult. Under these circumstances it is only possible to give a range of differential diagnoses.

Case 1: CIN III with viral changes

401

402

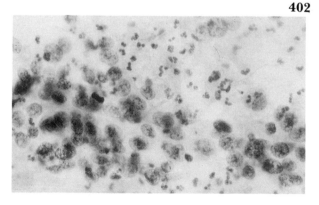

401 Dyskaryotic and keratinized cells.
Well-differentiated, squamous dyskaryotic cells are present with dyskaryotic columnar cells. Some keratinization is seen. (× *160*)

402 Undifferentiated dyskaryotic cells.
Elsewhere in the smear, clusters of poorly differentiated cells are found which have poorly defined cytoplasm. Note the irregular areas of clearing in some nuclei and the presence of nucleoli. This smear was reported as indicating at least CIN III with some features causing suspicion of invasion. (× *160*)

403

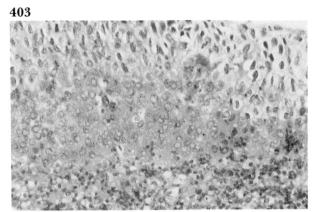

403 CIN III with viral changes: cone biopsy.
Sections from the cone biopsy showed CIN III only, but note the herpes-like changes in the nuclei of the basal layers. There is also a heavy inflammatory cell infiltration of the stroma, with some cells invading the epithelial layer. Single-cell keratinization is also present. (*H&E, × 80*)

Case 2: CIN II

404

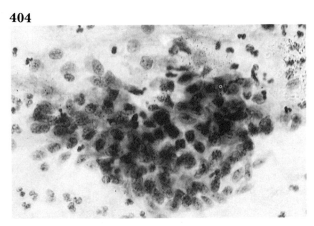

404 Undifferentiated dyskaryotic cells. The cervical smear from this patient showed a number of sheets and clusters of undifferentiated dyskaryotic cells. In this field the nuclear chromatin pattern is granular and a few single cells contain red nucleoli. (× *160*)

405

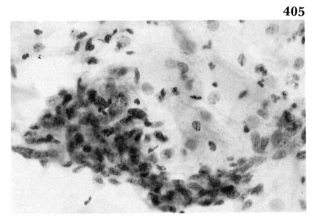

405 Undifferentiated dyskaryotic cells.
Elsewhere in the smear, this tissue fragment, which shows some fusiform nuclei, was found. (× *160*)

406

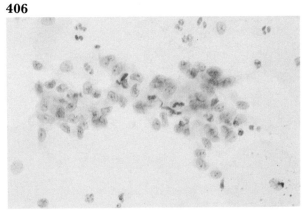

406 Dyskaryotic columnar cells. The cells in this cluster have paler nuclei but show prominent red nucleoli. (× *160*)

407

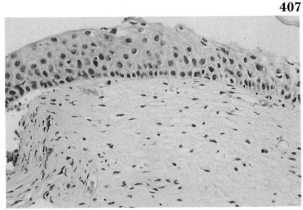

407 CIN II: colposcopic biopsy. Colposcopic biopsy showed CIN II only. (*H&E, × 80*)

408 CIN II: colposcopic biopsy. At higher magnification hyperchromatic nuclei, similar to those seen in the smear, are present but these show more cytoplasmic differentiation. Note that nucleoli are present in some cells. In view of the degree of abnormality in the smear, the biopsy report was unexpected. The discrepancy could result from failure to biopsy the most abnormal area, but in this case three smears taken during the following years were all negative. Regression following biopsy has been reported, and this can happen because the biopsy has removed the whole of the abnormal area or because the trauma of biopsy stimulates an immune response (*Armstrong et al., 1984*). (*H&E, × 160*)

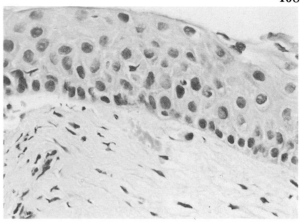

Case 3: Squamous cell carcinoma (early invasion)

409

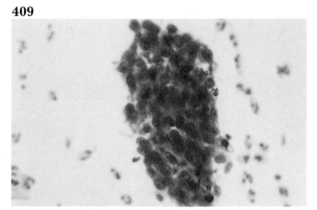

410

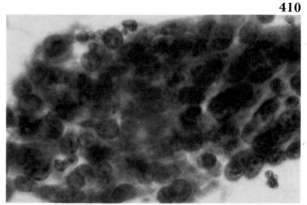

409 Poorly differentiated cells. This cluster of poorly differentiated cells was found in a routine smear. The nuclear chromatin is hyperchromatic and granular, with prominent chromocentres. (*× 125*)

410 Severe dyskaryosis. At higher magnification irregular clear areas are seen in the nuclei, together with prominent nucleoli and clearing around the nucleoli. The smear was reported as showing changes causing suspicion of an invasive lesion. (*× 250*)

411

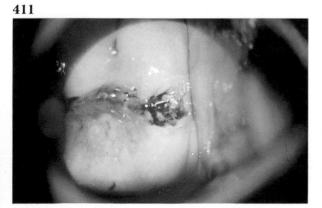

412

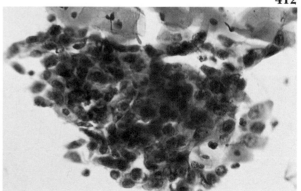

411 Colposcopy. There was some difficulty in tracing this patient, and she was not seen in the colposcopy clinic until 6 months later. Note the discrete area of vascular irregularity at the right angle of the external os. (*Saline, green filter, × 10*)

412 Severe dyskaryosis. The degree of cytological abnormality remained the same as at the time of the first smear. (*× 125*)

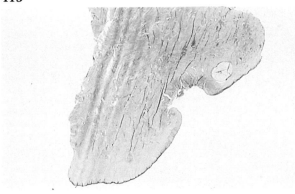

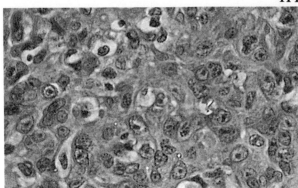

413 Early invasion: mounted section of cervix. This section was taken from the cone biopsy at the level of the area of vascular abnormality. It shows a small ulcer with early invasion at the base. (*H&E*)

414 Early invasion: cone biopsy. At higher magnification the cells at the invading edge of the tumour can be compared with those seen in the cervical smear. (*H&E, × 200*)

Case 4: Small cell squamous carcinoma

415

416

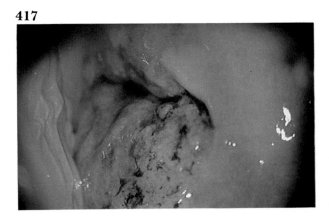

415 Undifferentiated dyskaryotic cells. At screening magnification the undifferentiated dyskaryotic cells in this field resemble those seen in the two previous cases. (*× 62*)

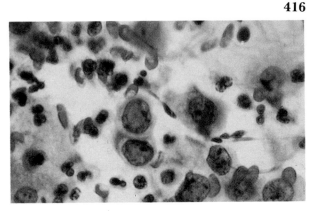

416 Malignant squamous cells. At higher magnification the nuclei are paler than in the previous cases and there is more extensive clearing and irregular condensation of nuclear chromatin, particularly at the nuclear membrane. Note also the angularities of the nuclear outline. (*× 250*)

417

417 Colposcopy. Compare the colposcopic photograph in this case with **411**. Note the granular area replacing the posterior lip of the cervix. (*Saline, green filter, × 16*)

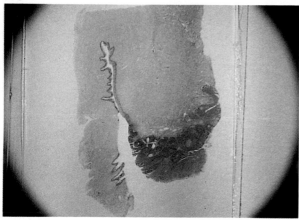

418 Section of cervix and vaginal wall. The patient was treated by Wertheim's hysterectomy. Note the plaque of tumour on the posterior lip of the cervix. (*H&E*)

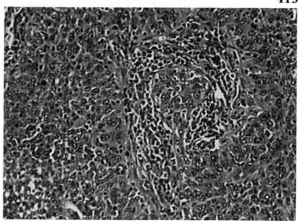

419 Small-cell squamous carcinoma: cervix. At higher magnification the cells in the tumour can be compared with those seen in the cervical smear. (*H&E, × 37.5*)

Case 5: Small-cell carcinoma

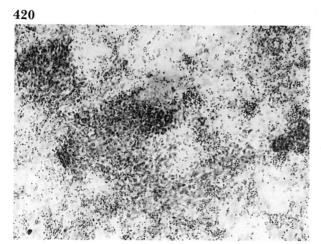

420 Undifferentiated dyskaryotic cells. At screening magnification there is a profuse exfoliation of undifferentiated dyskaryotic cells. (× *40*)

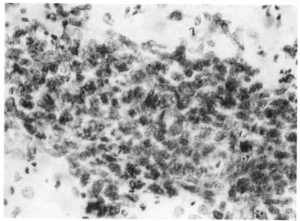

421 Undifferentiated malignant cells. At higher magnification most of the cells are in tight clusters with poorly defined cytoplasm. Irregularities of nuclear chromatin pattern suggest an invasive lesion. (× *160*)

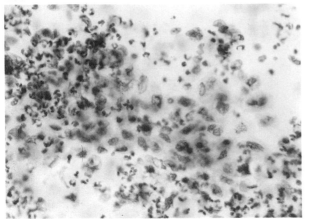

422 Cells with squamous differentiation.
Elsewhere in the smear some dyskaryotic cells show squamous differentiation. (× 160)

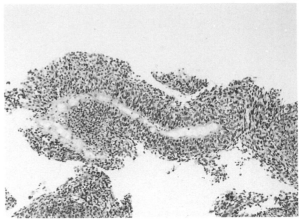

423 Small-cell carcinoma: biopsy of cervix.
This biopsy from the surface of the tumour shows strips of neoplastic epithelium. There is no stroma, so invasion could not be confirmed on this specimen, but clinically an overt tumour was present. (H&E, × 40)

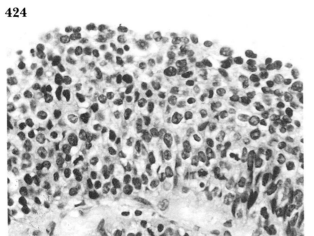

424 Small-cell carcinoma: biopsy of cervix. At the same magnification as **421** it can be seen that the cells making up the tissue in this section correlate with the predominant cells in the cervical smear. (H&E, × 160)

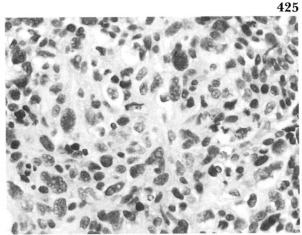

425 Squamoid change: biopsy of cervix.
Elsewhere in the section, areas of squamoid change are seen correlating with the cells in **422**. The histological diagnosis in this case was basi-squamous carcinoma. (H&E, × 160)

Case 6: CIN III with mucus secretion (courtesy Dr C.M. Stanbridge and Dr C.H. Buckley)

426

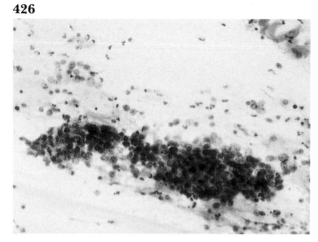

426 Undifferentiated dyskaryotic cells. This is another example of a patient presenting with dense fragments consisting of undifferentiated dyskaryotic cells with hyperchromatic nuclei. (× *80*)

427

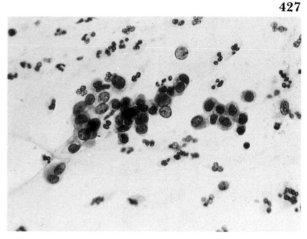

427 Severe dyskaryosis. Elsewhere in the smear, cells are seen in loose clusters permitting closer study of the morphology. The nuclei are hyperchromatic with a granular chromatin pattern. In some cells the cytoplasmic border is well defined, suggesting squamous differentiation, whereas in others the cytoplasm is flocculent and there is a suggestion of acinar formation. (× *160*)

428

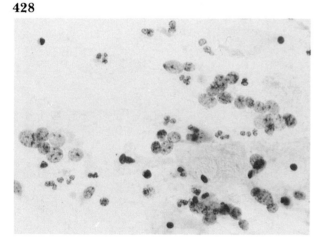

428 Bare nuclei. This field contains large, pale bare nuclei. The nuclear chromatin pattern is still granular and prominent nucleoli are present. (× *160*)

429

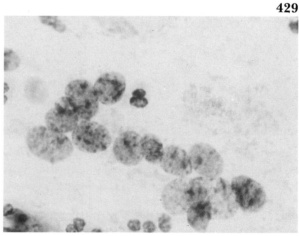

429 Bare nuclei. At higher magnification it can be seen that the nucleoli are red. The cytological features illustrated suggest a squamous and a glandular component to the lesion, and the possibility of one or both of these being invasive could not be excluded. (× *400*)

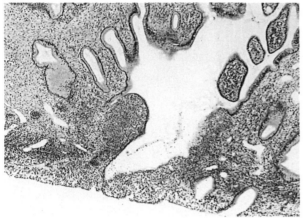

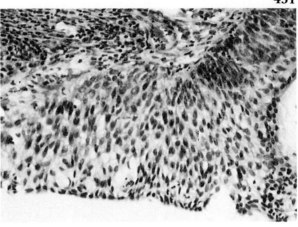

430 CIN III: cone biopsy. Examination of conventional sections from the cone biopsy showed CIN III with some gland involvement, but endocervical columnar epithelium lining the crypts was normal. (*H&E, × 16*)

431 CIN III: cone biopsy. This field shows part of the section at higher magnification. Note the columnar cells which are still present on the surface of the CIN III. (*H&E, × 80*)

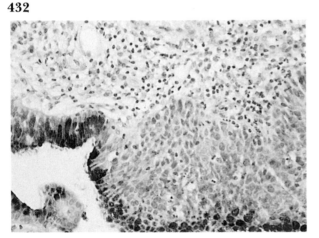

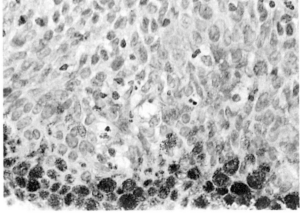

432 CIN III: cone biopsy. This section shows the transition between normal columnar epithelium and CIN III. In addition to positive staining for mucin in columnar cells, this is also seen at the surface of the CIN, and isolated cells in the depth of the epithelium show blue or pink staining. (*PAS–Alcian blue, × 80*)

433 CIN III (mucin-positive): cone biopsy. The isolated cells which stain for mucin are seen more clearly at higher magnification. Note the concentration at the luminal surface. (*PAS–Alcian blue, × 160*)

Case 7: Adenocarcinoma of endometrium

This last case is included as a reminder that adenocarcinoma of endometrium can also present with dense clusters of hyperchromatic cells.

434 Undifferentiated malignant cells. This field contrasts a cluster of malignant cells with a sheet of parabasal cells in an atrophic smear. The patient had had a previous hysterectomy for adenocarcinoma of endometrium. (× *80*)

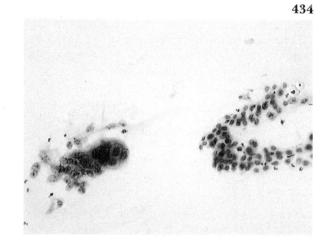

435 Undifferentiated malignant cells. Another field shows a sheet of undifferentiated malignant cells at higher magnification. (× *160*)

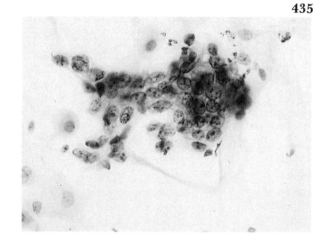

436 Adenocarcinoma of endometrium: hysterectomy. This patient had a vault recurrence which was clinically obvious, so no biopsy was taken. This section is from the original tumour, which was a well-differentiated adenocarcinoma. It should be remembered that recurrent and metastatic tumour is often less well-differentiated than the original. (*H&E*, × *40*)

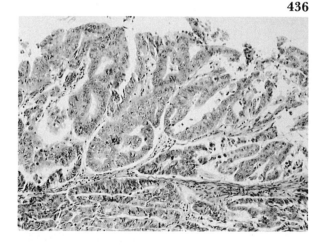

References

Armstrong D N, Butler E B, Stanbridge C M, Buckley C H. Correlation of colposcopic features and histological diagnosis in the cervix uteri. I. A retrospective study. *The Cervix and Lower Female Genital Tract*, 1984; **2**: 229–238.

Ayer B, Pacey F, Greenberg M, Bousefield L. The cytologic diagnosis of adenocarcinoma in situ of the cervix uteri and related lesions. I. Adenocarcinoma in situ. *Acta Cytologica* 1987; **31**: 397–411.

Buckley C H, Butler E B, Donnai P, Fouracres M, Fox H, Stanbridge C M. A fatal case of DES-associated clear cell adenocarcinoma of the vagina. *Journal of Obstetrics and Gynaecology* 1982; **3**: 126–127.

Buckley C H, Butler E B, Fox H. Cervical intraepithelial neoplasia. *Journal of Clinical Pathology* 1982; **35**: 1–13.

Buckley C H, Butler E B, Fox H. Vulvar intraepithelial neoplasia and microinvasive carcinoma of the vulva. *Journal of Clinical Pathology* 1984; **37**: 1201–1211.

Burghardt E. Pre-malignant conditions of the cervix. In *Clinics in Obstetrics & Gynaecology. Cancer of the Vulva, Vagina and Uterus*. Vol. 3, No.2. (Ed. F. A. Langley). Philadelphia: W B Saunders & Co., 1976: pp. 257–294.

Butler E B, Monahan P B, Warrell D W. Kuper brush in the diagnosis of endometrial lesions. *Lancet* 1971; **ii**: 1390–1392.

Butler E B, Taylor D S. The postnatal smear. *Acta Cytologica* 1973; **17**: 237–240.

Byrne P, Jordan J, Williams D R, Woodman C. Importance of negative result of cervical biopsy directed by colposcopy. *British Medical Journal* 1988; **296**: 172.

Choo Y-C, Anderson, D G. Neoplasms of the vagina following cervical carcinoma. *Gynecological Oncology* 1982; **14**: 125–132.

Coppleson M, Reid B L. *Preclinical Carcinoma of the Cervix Uteri*. Oxford: Pergamon Press, 1967.

DeWaard F, Baanders-van Halewijn E A. Cross-sectional data on estrogenic smears in postmenopausal population. *Acta Cytologica* 1969; **13**: 675–678.

DeWaard F, Pot B, Tonckens-Nanninga N E, Baanders-van Halewijn E A, Thijssen J H. Longitudinal studies on the phenomenon of postmenopausal estrogen production. *Acta Cytologica* 1972; **16**: 273–278.

Evans D M D, Hudson E A, Brown C L, Boddington M M, Hughes H E, Mackenzie E F, Marshall T. Terminology in gynaecological cytopathology: report of the Working Party of the British Society for Clinical Cytology. *Journal of Clinical Pathology* 1986; **39**: 933–944.

Fletcher S. Histopathology of papillomavirus infection of the cervix uteri: the history, taxonomy, nomenclature and reporting of koilocytic dysplasias. *Journal of Clinical Pathology* 1983; **36**: 616–624.

Frost J K. The cell in health and disease. *Monographs in Cytology*, Vol. 2. Basle: Karger, 1969: pp. 58–59.

Gardner B L, Dukes C D. Haemophilus vaginalis vaginitis: a newly defined specific infection previously classified 'non-specific' vaginitis. *American Journal of Obstetrics and Gynecology* 1955; **69**: 962–976.

Gupta P K, Hollander D H, Frost J K. Actinomyces in cervicovaginal smears: an association with IUCD smears. *Acta Cytologica* 1976; **20**: 295–297.

Gupta P K, Lee E F, Erozan Y S, Frost J K, Geddes S T, Donovan P A. Cytologic investigations in Chlamydia infection. *Acta Cytologica* 1979; **23**: 315–320.

Highman W J. Calcified bodies and the intrauterine device. *Acta Cytologica* 1971; **15**: 473–475.

International Committee on Histological Definitions *Acta Cytologica* 1962; **6**: 235–236.

Kaufman R, Koss L G, Kurman R J, *et al.* Statement of caution in the interpretation of papillomavirus-associated lesions of the epithelium of the uterine cervix. *Acta Cytologica* 1983; **27**: 107–108.

Levison M E, Trestman I, Quach R, Sladowski C, Floro C N. Quantitative bacteriology of the vaginal flora. *American Journal of Obstetrics & Gynecology* 1979; **133**: 139–144.

Macgregor J E. *Taking Uterine Cervical Smears*. British Society for Clinical Cytology, Aberdeen University Press, 1981.

Meisels A, Fortin R. Condylomatous lesions of the cervix and vagina. I. Cytologic patterns. *Acta Cytologica* 1976; **20**: 505–509.

Meisels A, Fortin R, Roy M. Condylomatous lesions of the cervix. II. Cytologic, colposcopic and histopathologic study. *Acta Cytologica* 1977; **21**: 379–390.

Miller A B, Barclay T H, Choi M G. A study of cancer, parity and age at first pregnancy. *Journal of Chronic Diseases* 1980; **33**: 595–605.

Morse A R. The value of endometrial aspiration in gynaecological practice. In *Advances in Clinical Cytology* (Eds Koss L G, Coleman D V). London: Butterworths, 1981: pp. 44–63.

Papanicolaou G N. *Atlas of Exfoliative Cytology*. Cambridge, Massachussetts: Commonwealth Fund, Havard University Press, 1954.

Patten S F. *Diagnostic Cytopathology of the Uterine Cervix*, 2nd ed. Monographs in Clinical Cytology No.3. Basle: Karger, 1978: p.65.

Prins R P, Morrow C F, Townsend D E, Disaia P J. Vaginal embryogensesis, estrogens and adenosis. *Obstetrics & Gynecology* 1976; **48**: 246–250.

Purola E, Savia E. Cytology of gynecologic condyloma acuminatum. *Acta Cytologica* 1977; **21**: 26–31.

Report of the Intercollegiate Working Party on Cervical Cytology Screening *Appendix II. British Society for Clinical Cytology: Recommended Code of Practice for Laboratories Providing a Cytopathology Service (1986)*. Royal College of Obstetrics and Gynaecologists, 1987: pp. 35–55.

Singer A, Jordan J A. The anatomy of the cervix. In *The Cervix* (Eds Jordan J A, Singer A.) Philadelphia: W B Saunders & Co., 1976: pp. 13–36.

Skaarland E. New concept in diagnostic endometrial cytology: diagnostic criteria based on composition and architecture of large tissue fragments in smears. *Journal of Clinical Pathology* 1986; **39**: 36–43.

Spriggs A I, Butler E B, Evans D M D, Grubb C, Husain O A N, Wachtel G E. Problems of cell nomenclature in cervical cytology smears. *Journal of Clinical Pathology* 1978; **31**: 1226–1227.

Stanbridge C M, Butler E B. Human papillomavirus infection of the lower female genital tract: association with multicentric neoplasia. *International Journal of Gynecological Pathology* 1983; **2**: 264–274.

Stanbridge C M, Mather J, Curry A, Butler E B. Demonstration of papilloma virus particles in cervical and vaginal scrape material: a report of 10 cases. *Journal of Clinical Pathology* 1981; **34**: 524–531.

Trevathan E, Layde P, Webster L. Cigarette smoking and dysplasia and carcinoma in situ of the uterine cervix. *Journal of the American Medical Association* 1983; **250**: 499–502.

van Niekerk W A. Cervical cytological abnormalities caused by folic acid deficiency. *Acta Cytologica* 1966; **10**: 67–73.

Ziabkowski T A, Naylor B. Cyanophilic bodies in cervico-vaginal smears. *Acta Cytologica* 1976; **20**: 340–342.

Index

The numbers in *italics* refer to pictures. Those in ordinary type to pages.